Political Epistemology

This series engages with the relation between knowledge and power. It focusses on historical, sociological and philosophical studies on science and knowledge in different forms. Thus, it aims to develop in particular the connection between historical-epistemological, philosophical and political perspectives in HPS, STS and other knowledge studies.

The central ambition of political epistemology, as we envision it in this series, is to trace and analyze the genealogy, validity, function, aims and conflictual constitution of knowledge and its consequences. Political epistemology, in this sense, does not only deal with questions of social validation of knowledge, in particular the relation between science, politics and truth, the critique of the organization and instrumentalization of science, but also expands historical epistemology, understood as a mode of reflection on scientific knowledge in a long-term global development. Political Epistemology takes up the historical, critical and comparative perspectives to focus on the entanglements of social and epistemic processes and to emphasize the materiality of knowledge as a world-transforming relation between society and nature on a global scale.

Wolfgang Lefèvre

Theory of Nature and Mode of Production

The Genesis of Modern Science in Materialist Perspective

Springer

Wolfgang Lefèvre
Philosophy and History of Science
Max Planck Institute for the History of Science
Berlin, Germany

Translated by
Cat Moir Wolfe
Toulouse, France

ISSN 3059-362X ISSN 3059-3638 (electronic)
Political Epistemology
ISBN 978-3-032-02294-3 ISBN 978-3-032-02295-0 (eBook)
https://doi.org/10.1007/978-3-032-02295-0

Translation from the German language edition: Lefèvre, Wolfgang. "Naturtheorie und Produktionsweise: Probleme einer materialistischen Wissenschaftsgeschichtsschreibung: eine Studie zur Genese der neuzeitlichen Naturwissenschaft." Darmstadt and Neuwied: Luchterhand, 1978.

This Springer imprint is published by the registered company Springer Nature Switzerland AG
The registered company address is: Gewerbestrasse 11, 6330 Cham, Switzerland

Introducing Political Epistemology: Editors' Remarks on the Series and the Book

Wolfgang Lefèvre's *Theory of Nature and Mode of Production*[1] inaugurates our new book series in Political Epistemology. We are particularly pleased to start it with the translation of a work which we regard as a classic in historico-political studies on science. This choice highlights a central ambition of political epistemology, as we envision it in this series, namely to trace and analyze the genealogy, validity, function, aims, and conflictual constitution of knowledge its consequences. Political epistemology, in this sense, deals not only with questions of social validation of knowledge, in particular the relation between science, politics, and truth, but also expands historical epistemology, understood as a mode of reflection on scientific knowledge along four axes of inquiry:

Genesis—the historical and socio-economic development of science and cognition;
validity—the intellectual and institutional legitimacy of knowledge claims;
function—the aims, social purposes, cultural politics, and ideological roles of science and knowledge;
and praxis—the materiality of knowledge as a world-transforming relation between society and nature in a global perspective.[2]

The connection between knowledge theory and political critique cannot be limited to the denunciation of biases, misinformation, and ideology. It must also address the historical formation of knowledge, the infrastructures, and conflicts that shape what counts as truth, and the material (even ecological) and symbolic conditions of knowing. It also must point to the material consequences of science as a force of production, reproduction, and transformation of realities—natural, social, and mental. If one wants to understand the role knowledge plays in this world and how it more specifically supports political decisions (in the stricter and the broader sense), one

[1] First published as: *Naturtheorie und Produktionsweise: Probleme einer materialistischen Wissenschaftsgeschichtsschreibung; eine Studie zur Genese der neuzeitlichen Naturwissenschaft* (Lefèvre 1978).

[2] See also Ienna and Omodeo, "An Italian Classic in Political Epistemology from the Seventies," in: Ciccotti, Cini et al. (2024), 9–50, esp. 31–42; Omodeo (2019); and Ienna (2025).

needs a critical theory informed by epistemology, which takes into account different systems of knowledge diachronically and synchronically. Since modern states and economies on a global scale rely on scientific labor and "expertise," it is a core task to analyze and critically engage with this mode of knowledge production in relation with others.

Theory of Nature and Mode of Production tackles these entwined dimensions of the emergence of modern scientific culture from a Marxist perspective. Its German subtitle stated that it deals with "problems of a materialist history of science" and it set out to investigate the emergence of modern science by bringing together different sources on the historical (especially capitalist) conditions of science, technology, and philosophy. This problematic is still fundamental to understanding the most recent metamorphoses of capitalism. In an age of populisms and anti-scientific propaganda ("post-truth"), philosophy of science and history of science gain even more relevance. Going beyond the binary opposition between culturalism and scientist naturalism, Lefèvre's work constitutes an instance of productive engagement with the question of the social function of science by offering not only historical explanations but also conceptual resources that remain valid as instruments for critique and emancipation.

The book was written in the tense political climate of post-68 West Berlin, at a time when the question of the relationship between science, society, and nature was once again gaining theoretical urgency. Yet, its meaning reaches beyond the local circumstances of the book's composition and first circulation. It addresses historical, systematic, and political questions in a time in which Radical Science Movements were getting momentum (Ienna and Freyberg 2023). It seems fair to say that political epistemology as a more general concern about the functions and goals of science and knowledge production emerged from this context, which also had its effects on the historiography of science.

Lefèvre analyzed the practical and ideological conditions of modern science drawing from historical-materialist assumptions and elaborated a position that programmatically opposed both idealistic (or "internalist") and technocratic conceptions of science. Instead of viewing natural theory, i.e. natural philosophy and science in the early modern sense, as an autonomous system of knowledge, he inquired into the conditions of its genesis and development—particularly in relation to mechanics and modern mathematical-experimental science. He combined this question with a materialist sociology, according to which modes of production are constitutive structures for the conception of nature in general (natural theory). The central thesis of the book is that modern natural science was not merely an intellectual revolution or a Promethean attempt at gaining power over nature, but emerged from the upheaval of the social relationship to nature ("social metabolism") in the context of bourgeois' interests.

Lefèvre develops this idea both historically and systematically by showing how specific forms of labor, technology, and social cooperation—in manufacturing, crafts, and the emerging exchange of commodities—created the conditions for a more abstract relationship to nature—in terms of theoretical understanding as well as in terms of practical interactions. In his book, nature is not regarded as an object of "pure knowledge," but as a socially shaped reality mediated by practices (labor and tools).

It is here expedient to stress some aspects of the contexts of this work and Lefèvre's biography.[3] The relevant *milieu* was West Berlin in the Sixties and Seventies. When Lefèvre arrived in Berlin as a young man, the wall had just been built. However, contact between both parts of Berlin was possible.[4] Lefèvre studied philosophy, history, sociology, and religious studies at the *Freie Universität Berlin* and became active in the *Sozialistischer Deutscher Studentenbund* (SDS, Socialist German Student Association), the intellectual and organizational backbone of the West German student movement. In 1965, he was elected head of the General Student Committee (AStA) alongside Peter Damerow—with whom he established a lasting intellectual alliance and developed a common research line of historical epistemology. They soon were dismissed from the AStA for their political positions, in particular their signature of a statement against the Vietnam War co-authored with colleagues from East Berlin—circumstances that reflect Cold War tensions and Lefèvre's commitment to an internationalist critique of power. He especially contributed to the socialist debunking of the economic interests behind education and research and described university as an apparatus that, under capitalist conditions, reproduces class hierarchies and subordinates scientific labor to entrepreneurship and technocratic agendas. The controversies on these topics culminated in the 1968 students' mobilization and the publication of *Die Rebellion der Studenten* (*The Students' Rebellion*), which Lefèvre wrote together with Rudi Dutschke and others to present a general vision of the anti-authoritarian APO (*Außerparlamentarische Opposition*, i.e., extra-parliamentary opposition). They argued that intellectual labor is intrinsically political and expressed their solidarity with "third-world" liberation movements.[5] After SDS had become a strong voice in a strained political arena that had witnessed the death of a student in 1967 and the attempt on Rudi Dutschke's life in 1968, the movement split up as radicalizing tendencies also came to the fore.[6]

This context of confrontation and repression is the background of *Theory of Nature and Mode of Production* as a work forged in struggle, marked by polemical urgency as well as theoretical rigor. It was animated by a commitment to mobilize knowledge and science for social transformations. Hence, Lefèvre began to part

[3] For more information, see the interview of Lefèvre (Grote et al. 2022).

[4] This contact proved very productive in the exchange with colleagues from East Berlin, especially from the Academy of Sciences, where in 1979 a new institute on Theory, History and Organisation of Science (ITW) was founded. At this time the socialist countries were often ahead in systematic science studies. See Kröber and Steiner (1972), Ruben (1976), Laitko (1996), and the famous study by Richta et al. (1968).

[5] Bergmann, Dutschke et al. (1968a). Giangiacomo Feltrinelli published an Italian translation in the same year (1968b).

[6] Rudi Dutschke (1940–1979), like some of the leading SDS members a refugee from the GDR, was shot on 11 April 1968 by a right-wing assailant, who was hyper-responsive to the political climate and the murder of Martin Luther King Jr. just the week before. Dutschke was hit twice in the head. Although he survived the attempt, the injuries forced him to mostly retreat from public and led to his premature death in 1979.

ways with both the most radicalized groupings and the romantic strains of the anti-authoritarian movement, which was influenced by the critique of science and society of the Frankfurt School (often reducing science to "instrumental reason"). Lefèvre's line of investigation engaged with the emancipative potential of education, knowledge, and the sciences and searched for the developmental potentials that emerge from within science.

At the end of his studies, during which he had been engaged in intellectual and practical struggles,[7] Lefèvre asked Jacob Taubes—an unorthodox professor, who connected Jewish theology and radical critique—to supervise his doctoral dissertation, which was devoted to an assessment of Weberian sociology. Despite Taubes' support, Lefèvre's dissertation met with severe institutional and professorial resistance. Four professors even attacked the validity of his dissertation in the effort to block his appointment to an academic position. They objected to his political standpoint and criticized his theoretical departure from rigid disciplinary norms. [8] The affair became emblematic of broader struggles over political dissent, pluralism, and academic freedom, which led to the infamous time of *Berufsverbote*, that is, the exclusion or even expulsion of left-wing intellectuals from West German universities for political reasons.

Eventually and in spite of opposition, thanks to Taubes and other colleagues at the *Freie Universität*, Lefèvre obtained the position of an assistant professor, a position defined as the study of the "history of science on a sociological basis." Lefèvre himself stated that, at that point, he was a newcomer to the field of the history of science.[9] The Marxist perspective he had already adopted, of course, provided him with instruments for such a new engagement. It should also be stressed that, in German-speaking universities, the history of science was not yet very much institutionalized at that time (Grote et al. 2022). *Theory of Nature and Mode of Production* presents the results of Lefèvre's entrance in the field. He intended to offer a rigorous as well as positioned account and an overview over the state of the more general discussion concerning the "Scientific Revolution." Since the contexts and networks of the German pre-war discussions on the topics were disrupted, the book represents, as it were, the Boris-Hessen-moment in German-speaking history of science. It showed that the connection of thorough historical research, theoretical reflexivity, and political commitment is possible. From this viewpoint, his work is still a reference point.

In Berlin, Lefèvre held a university position that combined the history of science with that of philosophy[10] and he brought this perspective along when he moved to

[7] Not only as a political activist but also as an organizer of projects like the planned alternative university at the FU campus (*Kritische Universität* in 1967/8).

[8] See the coverage especially in the famous German weekly *Der Spiegel* where several articles attest to the public attention. See in particular *Der Spiegel* (1971a, 1971b).

[9] See Grote et al. (2022) and the author's foreword (in this volume).

[10] During this time Lefèvre became engaged in the international Hegel Society and held a decades-long Hegel-Kolloquium with Peter Furth and Peter Damerow, which was attended by colleagues to-be like Peter McLaughlin, Hans-Jörg Rheinberger, and later also Jürgen Renn. Lefèvre com-

the Max Planck Institute for the History of Science, where he corroborated his research program with case studies on early modern science, technology, knowledge, and "the material means of thinking."[11] He thus continued his early commitment to a social history of science in new forms. Indeed, our idea to translate and (re-)publish his work of 1978 goes back to conversations we had with him at the Max Planck Institute for the History of Science nearly a decade ago. In a setting where historical epistemology served as that institute's guiding methodology—albeit according to different strands[12]—we reflected on how political concerns had been diluted in the studies on science, after scholarship and activism in the heated Sixties and Seventies had brought forward critical positions on science politics as well as on economic interests and knowledge.[13] In this respect Lefèvre's project was

bined insights of both spheres, e.g., in (1984) and in works on the science question in Leibniz, Kant, and Hegel, such as Damerow and Lefèvre (1983). The translation of the latter by Cat Moir is forthcoming in the online journal *Marxism & Sciences*.

[11] See also the programmatical account in Lefèvre (2000, 245–6): "If it is correct that scientific truth depends on a given material means of thinking, and if these means undergo historical change, then the crucial question arises of how the historical process of transformation from one set of means to another may be understood. The answer to this question is crucial for whether or not it is meaningful to speak of an historical development of science instead of considering this change a mere series of replacements of mutually indifferent approaches to knowledge. This concerns not only the process of transmitting systems of knowledge from one cultural tradition to another, but also the famous problem of incommensurability, which has been discussed by philosophers of science since Kuhn. Studying the replacement of material means by new ones in the course of history, we will find in most cases that the new ones allow us to perform tasks equivalent to the earlier ones. [...] The building of [a] new structure [of knowledge] included a reconstruction of the old one, based on the new conditions given by the new material means of thought. To the extent that this is an appropriate portrayal of the ways in which different traditions of thinking were integrated or the replacement of systems of knowledge took place in history, I think it is justified to speak of a truly historical development of science. We will then encounter not only continuity and coherence within such historical developments, but this also explains why and to what extent historical reconstructions of former systems of knowledge are possible at all. Furthermore, we may then be ab le to uncover certain regularities which govern such processes of integration and replacement. But surely, we will not find laws of development, which allow us to derive the development of science from inviolable principles or to predict the outcome of this development in the future. The historical development of science is a process that possesses both features: it is lawful and contingent at the same time."

[12] See the discussion in Feest and Sturm (2011).

[13] Cf. Lefèvre (2000, 239): "Since the nineteenth century, science played a crucial role in modern societies, not only in production and economics, but also in the field of politics. Politics builds on knowledge and insight, and this distinguishes political power from arbitrariness or violence. Gaining knowledge of a natural process that might be useful for purposes of production is a complex activity. However, to gain knowledge of social processes that might be useful for purposes of political decision is even more complex. Gaining knowledge of the social activity of science itself belongs to the latter category. Philosophy, history, and sociology of science do not provide a coherent picture of science. On the one hand, philosophy of science, despite all the scepticism about rationality in recent times, still tries to uncover the hidden thread of rationality within the scientific enterprise and its development. On the other hand, history and sociology of science show how science is embedded in and deeply dependent on changing social contexts so that we cannot any longer hope to uncover anything like a universal logic of the development of science. Given this

part of a large international movement for a critical Marxist perspective emphasizing emancipation in education, science, and technology. Education was a cornerstone of the '68 movements, when the ideological dimension of teaching first came to the fore as an important battlefield. In many countries, university reforms were aimed at mass education and student participation at the same time.

This context also favored a reappraisal of earlier studies on science, technology, and society, although punctuated with criticism. The legacy of Marxist philosophy and sociology of science was typically embodied by Boris Hessen's externalist historiography and John Bernal's analysis of the social functions of science. However, activists and scholars of the Seventies often distanced themselves from the perceived productivist optimism, which had dominated earlier visions of science as essentially emancipatory and at odds with capitalist power relations. By the Sixties, the non-neutrality of science, its antagonistic character, and the problem of ideology could not be neglected, anymore.

The engagement with these topics led students and scientists to denounce the non-neutrality of knowledge production. They questioned dominant science policies, especially the military and economic interests, exploitative mechanisms of technological progress, and technocratic ideology. Hilary Rose and Steven Rose's *Ideology of/in the natural sciences* (1976) is representative of the intersection of a wide range of themes ranging from socio-economic struggles, anti-capitalist criticism, feminist standpoints, and early environmentalism. In Rome, a group of physicists, Ciccotti, Cini, De Maria, and Jona-Lasinio, published a kind of manifesto *L'Ape e l'architetto* (*The Bee and the Architect*, 1977, English translation, 2024), which developed a historical-materialist critique of the non-neutrality of science.[14] Some of these works still stand out as critical self-reflections from within the sciences.

At the time when Lefèvre was still working his way into the field, the "Scientific Revolution" constituted a favored theme of historical and philosophical skirmish over science, its genesis, validity, and goals.[15] Lefèvre's work developed the

situation, it seems to make sense for an historian of science to look for elements of the scientific process that may serve as bridge-builders between these approaches. Candidates for such elements are what I will call the material means of thinking."

[14] See Ciccotti et al. (2024). From their historical materialist perspective, the notion of "scientific revolution" as influentially understood by Thomas S. Kuhn had to be explained through societal processes, including antagonisms and social revolution. Moreover, in line with Raniero Panzieri's proto-operaist reading of Karl Marx's *Grundrisse* (esp. the so-called *Fragment on Machines*), they reconsidered technological advance as the non-neutral result of capitalist investments for the sake of exploitation through the extraction of relative plus-value (via augmented productivity) and maximized control over the workers through automation. In the same years, Feminist socialists (among them, Selma James, Dalla Costa, Fortunati, Federici) expanded the critique of exploitation from the classical sites of production and class struggle (the factories) to broader societal relations, the household, and the biological premises of capitalism. They addressed the question of reproduction as functional to wage labor and the market, which constitutes a biopolitical aspect of relevance for the political epistemology of medicine and environmental studies.

[15] See also Lefèvre's remarks in his Foreword to the English edition (in this volume).

premises of early Marxist scholars of the Scientific Revolution, in particular Hessen and Borkenau: the former brought forward the idea that machine technology was at the basis of mechanical philosophy; the latter stressed the primacy of societal antagonisms (class struggle) over technology-centered histories of progress. Lefèvre expanded earlier contextual and historical materialist studies by pointing out the relevance of ideology, the "theory of nature" or natural philosophy, as a shaping factor of modern science inscribed in the class struggle of the bourgeoisie against feudal structures. Hence, Lefèvre renounced strict economicism.[16]

In its own Marxist way, the book prefigured the cultural turn of the Eighties, although it was opposed to mere relativism. The cultural turn in science studies in the long run produced a microhistory, in which the term "political epistemology" was rather used to point to political intervention, dogmatisms, or sociological conditions in very limited geographical, temporal, and institutional milieus. The publication of Steven Shapin and Simon Schaffer's *Leviathan and the Airpump: Hobbes, Boyle, and the Experimental Life* (1985) marked this turn, moving away from the political economy of science that had informed earlier (socialist) sociology of science. They referred to the motivations behind early modern British scientists as a problem in "political epistemology" as link between the conceptions of knowledge and political ideals of historical figures such as Thomas Hobbes (connecting rationalist philosophy and authoritarianism).[17] Soon after, the entrance of the postmodern in science studies explicitly rejected the "solid political epistemologies" rooted in class struggles, feminist Marxisms, and decolonial movements for relational and fluid pluralism.[18] Today's cultural and political conjuncture, marked by

[16] This was a common concern in the Marxist discussions of the 1960s and 1970s on both sides of the "iron curtain." Kindred concerns fueled different intellectual attempts to ground early modern philosophy in a progressive moment of capitalist bourgeoisie. A lot of these studies read local experiences against the background of broader historical, socio-economic, and ideological structures and processes. A famous example is Negri (1981/1991), a book which regarded Baruch Spinoza's radical ontology as the sign of the expanding revolutionary bourgeoisie in the seventeenth-century Netherlands, in a time in which the emerging class of modernity was forging its intellectual tools. The problem of the ideological abuses of science was tackled by leading figures of the Frankfurt School, in the USA and Germany, as in Marcuses' *One-Dimensional Man* (1964) and Habermas' *Technik und Wissenschaft als "Ideologie"* (1968, tr. *Technology and Science as Ideology*, 1970) and *Erkenntnis und Interesse* (1968, tr. *Knowledge and Human Interests*, 1987). British cultural studies emerged and radical science movements revised ideas stemming from French approaches (mainly structuralist), German philosophy (mainly linked to critical theory), and Italian philosophy of praxis (mainly Gramscian). In France Althusser and Foucault became prominent voices. For the development of the field of historical epistemology in the French context, see Ienna (2025). In the Eastern "bloc" the history of science and "science of science" was re-emerging in the 1960s (Russ. *Naukovedenie*, German: *Wissenschaftsforschung*) with the famous report of the Czech research group led by Radovan Richta, which proclaimed the end of the extensive phase of industrialization. Socialist projects like this had profound impact on the formation of groups like the "Club of Rome" and their concept of de-growth (Richta et al. 1969).

[17] See Shapin and Schaffer (1985, 99–107).

[18] Haraway (2016, 50–51): "Ambivalence toward the disrupted unities mediated by high-tech culture requires not sorting consciousness into categories of 'clear-sighted critique grounding a solid political epistemology' versus 'manipulated false consciousness,' but subtle understanding of

populist identity politics and propaganda, the limits of social constructivist relativism, and post-modern pluralism (the polycrisis of "post-truth" and "post-factual"), force us to look back at historical epistemological approaches that connected critique with positioned objectivity, historical thinking with anti-relativism, philosophical reflexivity with the rejection of self-referential speculative abstraction.

This English edition of *Naturtheorie und Produktionsweise* offers for the first time to non-German readers the 1978 work. Cat Moir had to fulfill the task of rendering a dense Hegelo-Marxist philosophical prose into a readable translation. We tried to support her in that as best as we could and to find the English translations of the cited literature. Together with Wolfgang Lefèvre we decided to complement the book with three shorter texts (appendices), which reflect further development of theses and further implications of the book.[19]

The conception of nature as a second nature, as systematically developed by Lefèvre in the first appendix "Second Nature and the Natural Sciences," an article published in 1979, but outlined while writing the larger work, forms the philosophical counterpart to the materialist-historical argument of the book. While the latter reconstructs the genesis of modern natural science from the historical upheavals of the early modern period and engages with the much-debated topic of the Scientific Revolution, the article on "Second Nature" reveals that this genesis was made possible by a profound transformation of the very concept of nature itself. As Lefèvre argues, philosophical reflection played a fundamental role in scientific and societal processes. Nature no longer appears as mere given or substance, but as a reality overlaid by social practice, by technology, labor, and symbolic representation—as nature that has passed through history. The connection between both texts lies in the thesis that natural science could only arise within the framework of a particular relationship to nature, shaped by labor and social mediation—and that this relationship itself is not "natural," but historical. "Second Nature" does not only mean a metaphor that reifies social relations as "natural." For Lefèvre, the expression "second nature" is often used to reify or "naturalize" social relations, but can also refer to the epistemological conditions of scientific knowledge. Nature, as it appears in science, is already selected, structured, and rationalized through social practice. In the second appendix, "Between Two Natures" (1992), Lefèvre furthers his investigation of the gap between first nature and second nature, offering a historically sensitive framework for an adequate theory of abstraction and reification. The synthesis offered in the third appendix, "Knowledge Systems in Historical Change" (Damerow and Lefèvre 1994) sets forth a comparative historical epistemology in

---

emerging pleasures, experiences, and powers with serious potential for changing the rules of the game." This should be read as a post-modern criticism of both Marxist and feminist epistemologies. Indeed, it is a criticism of socialist feminism in the name of perspectivism and fluid ontologies/identities.

[19] See also the author's "Foreword to the English translation" in this volume.

connection with cognitive sciences, a framework that tries to integrate different historical results based on the analysis of the material means of thinking.[20]

Re-reading Lefèvre along with the work of his colleague Peter Damerow led us to the question of why the approach of their early works, so deeply invested in political and historical questions, has become marginal in dominant strands of history and philosophy of science.[21] The political spirit of the book at hand, which stands out from "ideological antiquity" (in Alexander Kluge's expression),[22] allows us to extend our concern to a broader public of science-studies scholars who have been going through a new wave of social and political controversies over science.

Opening our series with *Theory of Nature and Mode of Production* also means emphasizing the relevance of theories of nature for epistemology, in line with the current material, "ontological," and political turn in science studies. In an age shaped by the Anthropocene—the concept that brought to the fore, in the Earth sciences and humanities, the techno-scientific transformation of the planetary system—the question of how we know nature is inherently political. The Anthropocene awareness also draws on earlier analyses on the transformative objectivity of knowledge alongside with its partiality, depending on the social conditions of scientific abstractions and political positioning. Lefèvre challenges the ideology of nature as passive and absolute, by showing that modern science was supported by the elaboration of new conceptions of nature, shaped by capitalist production and socially rooted symbolic forms of abstraction. In this perspective, historical materialism proves an appropriate means of grasping the historically mediated metabolism between knowledge, nature, and power.

We hope the series will foster different kinds of critical studies on knowledge in the emancipative spirit expressed by Franz Seiwert's print of 100 years ago, which we chose for the book covers: people stand in solidarity and in the back a flag proclaims: "knowledge [*Erkenntnis*] of the world leads to transformation of the world."

Hamburg, June 19, 2025
Sascha Freyberg and Pietro D. Omodeo (for the editorial collective)

---

[20] The latter topic was developed already in Damerow and Lefèvre (1981). Lefévre's programmatic article (2000) revisits these issues, providing a methodical discussion of how scientific knowledge emerges from historically structured practices.

[21] See also the questions posed by Renn (2020, 33): "While scientific and technical knowledge dominates our daily life, and while the survival of humanity in the Anthropocene depends on a thoughtful application of science-based solutions, the current mainstream of the history of science rarely contributes to these discussions. How can we change that? What kind of approach could do justice to a conception of science as a human practice involving, in an irreducible way, mental, material, and social dimensions? How can one conceive knowledge as being constrained, but not determined, by local and by larger political and economic structures? Moreover, what historical and political epistemology could help to restitute moral responsibility to the quest of science for knowledge?"

[22] See also Fredric Jameson's commentary on Kluge's "Notes from the Ideological Antiquity" as cinematic inquiry into *Capital* and the desire of Eisenstein to adapt *Capital* for the screen (Jameson 2009).

References

Bergmann, U., R. Dutschke, W. Lefèvre, and B. Rabehl. 1968a. *Die Rebellion der Studenten, oder Die neue Opposition*. Reinbek: Rowohlt.

Bergmann, U., R. Dutschke, W. Lefèvre, and B. Rabehl. 1968b. *La ribellione degli studenti ovvero La nuova opposizione*. Milano: Feltrinelli.

Ciccotti, G., M. Cini, M. De Maria, and G. Jona-Lasinio. 2024. *The bee and the architect: Scientific paradigms and historical materialism (VF02)*. Verum Factum: Venezia 2024. https://verumfactum.eu/volumes/the-bee-and-the-architect/.

Damerow, P., and W. Lefèvre. 1981. *Rechenstein, Experiment, Sprache. Historische Fallstudien zur Entstehung der exakten Wissenschaften*. Stuttgart: Klett-Cotta.

Damerow, P., and W. Lefèvre. 1983. Hegel und die weltgeschichtliche Einheit der Wissenschaft. In *Hegel-Jahrbuch 1983*, ed. W. R. Beyer, 38–48. Pahl-Rugenstein: Köln.

Damerow, P., and W. Lefèvre. 1994. *Wissensysteme im geschichtlichen Wandel*, MPIWG-Preprint series. Vol. 5. Hogrefe. https://www.mpiwg-berlin.mpg.de/sites/default/files/Preprints/P5.pdf.

Feest, U., and Th. Sturm. 2011. What (good) is historical epistemology? Editors' introduction. *Erkenntnis* 75:285–302. https://doi.org/10.1007/s10670-011-9345-4.

Grote, M., A. te Heesen, and W. Lefèvre. 2022. Bausteine zu einer Oral History der Wissenschaftsgeschichte. Wissenschaft als Arbeitsprozess. Interview mit Wolfgang Lefèvre. *Berichte zur Wissenschaftsgeschichte* 45 (1–2): 265–280. https://onlinelibrary.wiley.com/doi/10.1002/bewi.202200013.

Habermas, J. 1970. Technology and science as 'ideology'. In *Toward a rational society: Student protest, science, and politics*, ed. J. Habermas, 81–122. Boston: Beacon Press.

Habermas, J. 1987. *Knowledge and human interests*. Boston: Polity Press.

Haraway, D. 2016. *A Cyborg Manifesto: Science, technology and social-feminism in the late twentieth century*. Minneapolis: University of Minnesota Press. (first published 1985).

Ienna, G. 2025. *Genesis and development of French historical epistemology: A trajectory toward political epistemology*. Transl. R.E. Burket. Cham: Springer-Nature.

Ienna, G., and S. Freyberg. 2023. Radical science movements: Past, present and future. *Marxism & Sciences* 2 (2): iii–xiv. https://doi.org/10.56063/MS.2301.02100. https://marxismandsciences.org/radical-science-movements-past-present-and-future/.

Jameson, F. 2009. Marx and Montage. *New Left Review* 58 (2): 109–117.

Kröber, G., and H. Steiner. 1972. *Wissenschaft: Studien zu ihrer Geschichte, Theorie und Organisation*. Berlin: Akademie.

Laitko, H. 1996. Wissenschaftlich-technische Revolution: Akzente des Konzepts in Wissenschaft und Ideologie der DDR. *Utopie kreativ* 73–74:33–50.

Lefèvre, W. 1978. *Naturtheorie und Produktionsweise: Probleme einer materialistischen Wissenschaftsgeschichtsschreibung; eine Studie zur Genese der neuzeitlichen Naturwissenschaft*. Neuwied: Luchterhand.

Lefèvre, W. 1979. Die zweite Natur und die Naturwissenschaften. Überlegungen zu den Aufgaben der Philosophie in der Diskussion um die Grundlagen der Naturwissenschaften. *Philosophia Naturalis* 17:242–253.

Lefèvre, W. 1984. *Die Entstehung der biologischen Evolutionstheorie*. Frankfurt am Main: Suhrkamp.

Lefèvre, W. 1992. Zwischen zwei Naturen. In *Der Comenius-Garten — Eine Leseprobe aus dem Buch der Natur*, ed. H. Vierck, 49–53. Berlin: Hentrich.

Lefèvre, W. 2000. Material and social conditions in a historical epistemology of scientific thinking. In *Science and power: The historical foundations of research policies in Europe*, ed. Luca Guzzetti, 239–246. Luxembourg: European Communities.

Marcuse, H. 1964. *One-dimensional man: Studies in the ideology of advanced industrial society*. Boston: Beacon Press.

Negri, A. 1991. *The savage anomaly: The power of Spinoza's metaphysics and politics*, trans. by M. Hardt. Minneapolis: Minnesota University Press (first published 1981 as *L'anomalia selvaggia* by Feltrinelli).

Omodeo, P. D. 2019. *Political epistemology: The problem of ideology in science studies*. Cham: Springer-Nature.

Renn, J. 2020. *The evolution of knowledge: Rethinking science for the anthropocene*. Princeton University Press: Princeton.

Richta, R. et al. 1969. *Civilization at the crossroads: The social and human context of the scientific and technical revolution* (1st ed. in Czech 1966). White Plains, NY: International Arts and Sciences Press.

Rose, H., and S. Rose. 1976. *The political economy of science: Ideology of/in the natural sciences*. New York: MacMillan.

Ruben, P. 1976/2010. Wissenschaft als allgemeine Arbeit. Über Grundfragen der marxistisch-leninistischen Wissenschaftsauffassung. *SOPO* 1976 (8/2): 7–40. Republished in: *Philosophische Schriften*, ed. U. Hedtke and C. Warnke. http://www.peter-ruben.de/schriften/Grundlagen/Ruben%20-%20Wissenschaft%20als%20allgemeine%20Arbeit.pdf.

Shapin, S., and S. Schaffer. 1985. *Leviathan and the Airpump: Hobbes, Boyle, and the experimental life*. Princeton and Oxford: Princeton University Press.

N.A. 1971a. Yankee am Hof. *Der Spiegel* 3: 51–53. https://www.spiegel.de/politik/yankee-am-hof-a-e627651f-0002-0001-0000-000043375212.

N.A. 1971b. Integriert sind wir nicht. *Der Spiegel* 3. https://www.spiegel.de/politik/integriert-sind-wir-nicht-a-0f44c02a-0002-0001-0000-000043375214.

Preface to the English Edition

The book *Naturtheorie und Produktionsweise* was published more than 40 years ago. At that time, the so-called "Scientific Revolution" was at the center of interests in the historiography of science, then a young and not well-established academic discipline. Most of the disputes of the historians of science revolved explicitly or implicitly around the question of how to explain and reconstruct the emergence and early development of the modern natural sciences in the West. The book was intended as a contribution to these disputes, though without much effect. The latter was due not least to the fact that it was written in German; it remained largely unknown in the tone-setting community of English-speaking historians of science. Pietro Daniel Omodeo's and Sascha Freyberg's proposal to publish an English translation of the book therefore came as a most welcome surprise and I am extremely grateful to them as well as to Catherine Moir, the translator.

The question naturally arises whether a translation of a text that was a contribution to an old dispute makes any sense at all. Today, the fierce disputes about the "Scientific Revolution" seem to belong to a bygone age that might be of historical interest only. Therefore—before addressing some aspects of the book—a few additional remarks regarding these old disputes may be in order, not least since reminding this bygone stage of the historiography of science might shed new light on this discipline in the past and present.

***

The topic of the "Scientific Revolution" has never been a purely historical one. Rather, from the very beginning it was ideologically charged. Taking Pierre Duhem's *Le système du monde* as a starting point, it centered initially around the question of whether or not the view of a sharp break between scholastic natural philosophy and modern natural sciences is tenable and, along with this, the widespread view of a principal antagonism between the new sciences and the Christian dogmas, in particular the catholic church and its claim to truth. Thus, with the topic of the "Scientific Revolution" the historiography of science was unavoidably entangled in ideological controversies.

Since the 1930s, triggered by the famous session *Science at the Crossroads* in the 2nd International Congress of the History of Science in London, the controversies about the Scientific Revolution were re-centered around a new problem: the question of how to reconstruct the emergence and development of the new, modern sciences. Was this to be done in the way of a history of ideas or in that of a social history? It may suffice to recall Alexandre Koyré and Alfred Rupert Hall as regards the former approach and, as regards the latter, Boris Hessen, Henryk Grossman, and Edgar Zilsel. As with the question of whether there was a break or continuity between medieval and early modern science, this new controversy was not purely a professional debate. Rather it was part and parcel of the age-old struggle between idealistic and materialistic philosophical convictions which took on a specific political character within the "system confrontation" between the capitalist and communist camps. This political character was even further reinforced in the Cold War after the Second World War.

The disputes at that time about the Scientific Revolution concerned not only professional or methodological questions entangled in political struggles. They were at the same time debates about the self-image of the modern industrialized societies that emerged from the *Siècle des lumières* and the early modern political revolutions. Science stood for rationality, technological and social progress, or the advanced industrial mode of production. *Nota bene*, both, West and East considered themselves inheritors of this historical legacy. In these ideological battles, Rupert Hall's concept of "modern scientific attitude" embracing scientific values such as rationality, disinterested search for truth and objectivity, was considered a signum of the West that explained and justified its alleged superiority and its factual domination of all other civilizations. It was this constellation that rendered the history of science a new medium of social reflection and, thus, a discipline that enjoyed public interest and appreciation. The history of science's boost and establishment as an academic discipline after 1945 must be seen in this context.

At the same time, after the two World Wars, and particularly after Hiroshima and Nagasaki, another perception was forming that considered science primarily as a key tool of advanced technologies of destruction and, a few decades later, of devastation of the natural environment with dangerous consequences for the climate. The generally affirmative, uncritical view of the sciences characteristic of the disputes about the Scientific Revolution had only temporarily pushed into the background, not replaced, the skepticism about science that originated in the wake of the First World War and remained an undercurrent of the perception of the sciences ever since. Just remember Edmund Husserl's *Krisis* essay of 1936 (Husserl 1970) or Jürgen Habermas' *Technik und Wissenschaft als Ideologie* of 1968 (Habermas 1970) in Germany or, in France, with different accents, Michel Foucault's and Jacques Derrida's writings. With this resurgence of skepticism toward claims to rationality in general and science in particular, ended the debate about the Scientific Revolution—or more precisely—this debate became an issue of a special historical discipline that was not of much interest beyond the professional boundaries. The heyday of the history of science as a medium of social reflection and self-understanding of the industrialized "First World" was over.

What ensued was again ideologically charged debates—for example, about science as ideology, as a means of suppressing other forms of world understanding, of suppressing artisanal accomplishments and alternative techniques or artistic creativity. But these new debates also included efforts to understand the tremendous importance of modern sciences in today's developed societies—key term: *Technoscience*. These efforts' focus was no longer on the origin and emblematic significance of the modern sciences but rather on their practical role in the highly developed industrialized societies; once more some key terms: science as a subsystem of the social production process, science—as Marx conceived it—as "factor" or "function" of this process in which science "appears as a force belonging to capital vis-à-vis labor" (MEW 26/1, 367f.). In these new debates, historiography of science has no privileged voice. On the contrary, it seems that historiography of science gets increasingly absorbed and replaced by cultural-historical, media-scientific, etc. discourses. But this is a different topic. Here it is about what does the present book, written in the era of the disputes about the Scientific Revolution, still have to offer for our understanding of the development of the modern sciences in the early modern period.

***

Naturtheorie (theory of nature) and *Produktionsweise* (mode of production)—these terms or concepts have a clear meaning in relation to the industrialized societies that developed since the turn of the eighteenth century. "Theory of nature" here stands for the modern natural sciences and "mode of production" for the capitalist industrial production. However, in relation to the early modern period of the European West, meaning and reference of these terms are not that clear.

As regards *theory of nature*, one encounters, in this period, a situation in which various speculative philosophical systems of nature existed side by side with mathematical conceptualizations of certain areas of natural phenomena, such as of statics, kinematics, or dynamics, and furthermore with systematic naturalistic descriptions and classifications of the kingdoms of minerals, plants, and animals. Needless to say, it would be unacceptable to consider only those theories that meet the standards of today's philosophy of science. Rather, the challenge is an adequate understanding of these different ways of thinking or beliefs about nature, of the preconditions and foundations of these ways of thinking as well as their interrelations.

As regards *mode of production*, one encounters yet another complex situation. As is well known, Marx's concept of mode of production denotes the unity of productive forces and social relations of production in a specific historical situation, that is, the specific unity of the material side of the production process (materials, tools, techniques, etc.) and the relationships between the social classes involved. As for the productive forces, alongside pre-industrial agricultural production, which was the dominant sector, there was a sector of handicraft production in the cities, as well as places with manufacturing or even incipient industrial production. Similarly, in terms of social relations of production, alongside the dominant relationship between peasants and feudal lords, there were social relations within the middle and bourgeois classes. These included relationships between masters and apprentices in trades and crafts, as well as between wage workers and capitalists in cottage industries, manufactories, and factories.

Finally, as regards the connection between the *view of nature* and *mode of production,* it should not be forgotten that men's relationship to nature is primarily a practical and not a theoretical one. Their understandings of nature develop in and with their practical relationship with nature, that is, develop in accordance with the forms of production through which they realize this relationship. Both, the productive forces and the social relations of production shape basic understandings of nature. Different views of nature thus correspond to different forms of labor. In preindustrial agricultural societies, nature is generally perceived as a life-giving, maternal power that humans can only cultivate and foster. From the point of view of artisanal production, nature does not appear as a life-giving power but rather as an inexhaustible arsenal of materials which can be exploited and transformed by human labor into consumer goods and the tools of their fabrication. Finally, on the basis of a predominantly industrial production, which arranges materials so that natural potencies, mechanical, electrical, chemical, etc., act as productive forces, nature can be and is conceived as a whole of mutually acting potencies. Different social relations of production, class relationships also play a role for the understanding of nature, albeit in a more indirect way, namely mediated through overarching understandings of the world, that is, mediated by understandings we encounter in religious and philosophical systems—key terms: monistic/dualistic or materialistic/idealistic, etc. worldviews. Regarding this extremely complex situation in which all of these different factors interact with each other, constantly changing their relative importance and readjusting their balance—how can a reconstruction do justice to all of these entanglements?

***

I do not pretend to have been fully aware of the complexity of my subject when I wrote my book back in the 1970s. As a professional philosopher who focused on the history of philosophy and had adopted a Marxist approach to history, three aspects of the genesis of the modern natural sciences in the early modern period seemed to me to be in particular need of clarification. As can be expected of a Marxist author, the *first* of these aspects concerned the question of the connection between the developments of the various production sectors of the time and the beginnings of modern natural sciences. Fully aware of the pitfalls of technological determinism, I showed that there was almost no need for scientific knowledge in any of these production sectors (at least not until the end of the seventeenth century). However, developments in some technologically advanced production sectors, such as mining, metallurgy, and ballistics, proved indeed conducive to the development of natural sciences—not only as offering challenging problems, but above all as technical prerequisites for experimental research.

The *second* aspect concerned the question of the significance and the role of the incipient natural sciences for and in the ideological struggle between the feudal and the bourgeois classes—or, in philosophical terms—for and in the struggle between a Christian-Platonist worldview and one that saw the appearing world as founded in itself.

Finally, the *third* aspect concerned the immaturity and contradictions of the new sciences, especially as regards new speculations about nature, such as philosophical atomism or mechanism. Further, as can be expected of a historian of philosophy, I pursued the question of how these contradictions were reflected in the philosophical systems of the seventeenth and eighteenth centuries, since the efforts in natural science and philosophy were closely intertwined at that time—a point that is unfortunately too often neglected by both historians of science and historians of philosophy.

An adequate understanding of the complexities, coincidences, and contradictions of this historical process still is a desideratum. I see my book just as an *al fresco* attempt at a better and more differentiated understanding of this process, and I hope that its new English edition will still be useful for historians working on this early stage of modern science.

***

At the suggestion of Pietro Omodeo and Sascha Freyberg, an appendix offers translations of three short articles, dealing with certain overarching aspects of the historical development of views and conceptions of nature that were considered but not explicitly addressed in *Naturtheorie und Produktionsweise* of 1978.

Two of these articles—*Second Nature and the Natural Sciences* (1979) and *Between two Natures* (1992)—discuss the correlation between practical relations with nature and the views, images, and concepts of nature in the major historical stages of Homo sapiens' practical relations with nature—beginning with the prehistoric stage in which human societies subsisted in nature as hunters and gatherers, followed by the stage of societies of herdsmen and farmers, which cultivated a producing nature, then the stage in which this cultivating way of producing was accompanied and supplemented by an increasingly differentiated artisanal production that saw nature as an inexhaustible arsenal of materials, and finally the stage of industrial production, in which the first nature was joined by a second nature constellated by man.

A further article, co-authored with the late Peter Damerow—*Knowledge systems in historical change* (1994)—deals with other aspects of the developmental stages of pre-scientific and scientific thought that were only marginally, but not explicitly, dealt with in the 1978 book. These include the historical development of knowledge systems depending on the historical development of cultural (that is, external as opposed to internal/mental) representations—oral language, writing, symbolic representations (e.g. number systems), etc.—as well as the historical development of forms of intellectual work.

Berlin, Germany
April 2025

Wolfgang Lefèvre

References

Damerow, P. and Lefèvre, W. (1994). *Wissensysteme im geschichtlichen Wandel*, MPIWG-Preprint series vol. 5. https://www.mpiwg-berlin.mpg.de/sites/default/files/Preprints/P5.pdf

Habermas, J. 1970. Technology and science as 'ideology'. In *Toward a rational society: Student protest, science, and politics*, ed. Habermas, 81–122. Boston: Beacon Press.

Husserl, E. 1970. *The crisis of European sciences and transcendental phenomenology*, trans. D. Carr. Evanston: Northwestern University Press.

Lefèvre, W. 1979. Die zweite Natur und die Naturwissenschaften. Überlegungen zu den Aufgaben der Philosophie in der Diskussion um die Grundlagen der Naturwissenschaften. *Philosophia Naturalis* 17:242–253.

Lefèvre, W. 1992. Zwischen zwei Naturen. In *Der Comenius-Garten — Eine Leseprobe aus dem Buch der Natur*, ed. H. Vierck, 49–53. Berlin: Hentrich.

Preface to the 1978 German Edition

The present study was written while I was working at the Institute of Philosophy at the Free University of Berlin, where I held an assistant position in the field defined as "history of science on a socio-historical basis" [*Wissenschaftsgeschichte auf sozialgeschichtlicher Grundlage*].

The fact that at the end of the 1960s some posts at the Institute of Philosophy of the Free University of Berlin (FU) were given a designated focus on history of science was certainly a far-sighted consideration in a scientific line of enquiry that is only now, in the mid-1970s, gaining general interest and recognition. However, this does not apply equally to the specification "on a socio-historical basis" given to the duties of my assistantship. The claim that—notwithstanding a few important scholars in England—the socio-historical treatment of the history of science in the Western world belongs to the 1920s is unlikely to provoke contradiction; it is—at least in terms of actual research work—"over," even if some works by veterans of socio-historically oriented historiography of science such as B. Hessen, F. Borkenau, or H. Mineur are still reprinted today. Now, although nostalgia is no basis for legitimizing scientific work, we must nevertheless reflect on the reasons why the socio-historical orientation in the history of science is now seen as obsolete, and whether—and if so, how—we might be justified in picking up this dropped thread again.

It is already well known that the research method in the history of science, insofar as it goes beyond philology, has hardly been thoroughly discussed, let alone developed. Moreover, if we look at the theoretical context of the socio-historically oriented history of science of the 1920s, it is clear that the perception of the field as outdated is not due to methodological progress made in the meantime. It would therefore be pointless to search in this direction for the reasons why the field is now seen as passé. This context was marked by Max Weber's studies on "occidental rationality," by Freud's essay *Civilization and its Discontents* (*Das Unbehagen in der Kultur*), by Husserl's *Krisis* lectures; the Critical Theory of the Frankfurt School as well as Heidegger's critique of Descartes belong to it as well. What makes this spectrum of theories—which, by theoretical and political standards, can hardly be surpassed in terms of "pluralism"—into a context from the point of view of interest here is a certain mistrust, common to them all in principle, of the scientific spirit, which—apart from Nietzsche and the conservative cultural critics—prior to 1914, Europe saw as a sign justifying its position of world domination. To those theorists, this scientific spirit no longer rhymed with progress or even humane civilization, but rather—at least to some of them—with poison gas or tanks. Weber and Freud registered the catastrophic antagonism of "rationalization" as a fateful doom; Husserl and Heidegger, but to a certain extent also Horkheimer, on the other hand, sought to show a way out of the impasse by tracing the historical path back to its origins. The history of science was thus challenged as never before.

In this history, the crisis of confidence in modern civilization and its rationality found no less "pluralistic" expression than in the range of theories outlined above. As if the two components of Weber's work on Protestantism were to be set against one another, two tendencies emerged in the history of science at this time. First, in light of the—certainly worthy—investigation of its prehistory in the Middle Ages, modern science was relativized as only one scientific theory among a diffuse set of ultimately equally irrational theories. Meanwhile, the socio-historically oriented history of science also took shape. For both directions (for the first of which the names Duhem, Koyré or Michalski are representative, and for the second Borkenau, E. Conze, Olschki, or Zilsel), it is true that modern science can no longer be regarded as "the" science whose genesis is "explained" by telling the story of the overcoming of prejudices and errors, of bias and faith in authority, and tracing the gradual accumulation of "correct" knowledge (however widespread this kind of history remained). However, the "progress" of the relativizing direction in relation to this pseudo-history consists in the fact that it regresses the history of science to a mere intellectual-historical description, to ideography, while the socio-historically oriented direction, with its attempt to make the history of science a truly explanatory science, actually points beyond that narrative history—however successful or unsuccessful this attempt may turn out to be in individual cases.

Clearly, the development of the history of science into an explanatory science depends on whether interest in such an explanation arises from problems that the respective present has with science. Thus, in order to explain the seeming outdatedness of socio-historically oriented historiography of science, we must ask why such an explanation is apparently little in demand today. In answering this question, as complex as any answer must be, the fact can hardly be overestimated that at the beginning of our century, bourgeois civilization became aware of its crisis internally, as it were, while today it is in competition with a civilization that claims to have recognized and remedied the causes of that crisis. How else can we understand that the reaction to Hiroshima and Nagasaki was less marked than that of the 1920s? The shock that marked the self-image of science in the period around the First World War must therefore be regarded as having been repressed rather than dealt with. Because the socio-historical orientation in the history of science, which emerged from this shock, was repressed along with it, I believe it is justified to resume this approach. Yet this endeavor also involves reviving the claim to pursue the history of science as an explanatory science. While it may not be possible to reach agreement on the question of whether an explanatory history of science is even conceivable, it may be possible to agree on the fact that the preoccupations that emerged from the philosophy of science in the 1960s represent no alternative to the socio-historically oriented approaches in this respect. Kuhn's "paradigm shift," as has become well known, does not explain, but is at best an aid to organizing the historical material.

Given this situation, I became convinced that it is legitimate and reasonable to pursue "history of science on a socio-historical basis" according to the definition of my post. However, if the socio-historical orientation is to be resumed, reflecting on its historical conditions also raises the question of whether it would be problematic to replace the "classical" topic with which this approach has been concerned—the genesis of modern science—with something else. As much as the interest in this topic may have sprung from the questionable hope of discovering, at the birth of the modern era, possible reactivatable alternatives to the development that actually took place, it nevertheless unmistakably addresses the central question of the relationship between scientific and social development in a way that, in my opinion, is incomparably suitable for working toward an explanatory history of science. For, to question the genesis of modern science is also to question the genesis of the form of the relationship between scientific and social development, whereby the apparently natural character [*Naturwüchsigkeit*] of this relationship is dissolved and it can thus itself be approached scientifically. A further argument in favor of this "classical" topic is that the discussion of the socio-historically oriented approaches in the historiography of science, with which this study is primarily concerned, can be conducted more fruitfully than in a debate on principles. The placement of my work in the program of an institute for philosophy also points in the same direction.

To consider the development of philosophy in its relation to the development of the sciences undoubtedly means more than just a welcome broadening of the horizons of research in the history of philosophy. Such an enterprise rather concerns philosophy at its core, concerns it as "first science," as founding theory, concerns it in its autonomy. A few comments to those who would argue that this even means pursuing the "abolition of philosophy" anew. However the "abolition of philosophy" may be judged, it is a process that has been taking place for more than two hundred years. Even the most resolute defender of *prima philosophia* cannot ignore this process or, in the posture of the armchair philosopher, interpret it as "decay" or as the work of agents hostile to philosophy; precisely, he is required to prove the reality of philosophy by facing this actual process.

The "abolition" of philosophy in the sense of *Aufhebung* is undoubtedly closely connected with the separation of former philosophical disciplines from philosophy and their establishment as independent sciences. This process begins with the genesis of the modern natural sciences. To see this process as natural implies that philosophy simply continues to play its role as a founding theory even though the scope of its intellectual task has been reduced. As a result, philosophy is often met with the contempt due to a theory that seeks presumptuously to explain to the individual sciences the state of the art of their own theoretical principles as if philosophy still provided the fundamental framework—an endeavor that, given the historical contingency of the state of the art, can at best be mere duplication if it is not to be exposed to future embarrassment. Yet, philosophy would only do justice to its tradition, to the Promethean impulse to which it owes its origin, if it seeks to comprehend the concealed history of what happens to the individual sciences, as to itself, in this "natural" process. In this light, the study of the history of science acquires a genuinely philosophical character, since investigating the moment when the individual sciences and philosophy became autonomous projects is undoubtedly of special importance.

Berlin, Germany Wolfgang Lefèvre

Contents

Chapter 1
The Development of Material Production and the Genesis of Modern Natural Science

1.1 First Section

1.1.1 When and Where?

If the only task of a materialist historiography of science were to establish the dependence of the development of science on economic and social developments, it would hardly have to reckon with problems related to the genesis of so-called classical, i.e. modern, natural science, which first achieved a state of internal coherence in the form of seventeenth-century physics. The connection between this classical natural science and the emergence of the bourgeois mode of production is so obvious that it can already be demonstrated to a certain extent in the temporal and spatial coordinates of this development.

Even the beginnings of this scientific development, which can be traced back to the High Middle Ages, point to centres of the urban bourgeoisie such as northern France and southern England, where the bourgeoisie had already become a factor in national politics in the thirteenth century, thus constituting national politics in the first place. In these regions we find not only scholars and veritable research scientists such as Grosseteste, Roger Bacon, Buridan and Nicole Oresme, but also the philosophical pioneers of a new worldview and thus also of nature: figures such as Roscellin, Abelard, Siger of Brabant and Ockham.[1] The dominant cultural and scientific position of northern Italy from the fourteenth to the sixteenth century was accompanied by an economic flourishing owed to its position in the Levant trade, which many economic

[1]Toledo, which perhaps played the most important role in the recovery of the scientific achievements of antiquity and the Arab bourgeoisie, may be disregarded here, although Alfonso X's cultural policy also seems to me inconceivable without the political role that accrued to the Spanish bourgeoisie through the liberation of Spanish territory. The situation in Frederick II's Sicily is more complicated.

W. Lefèvre, *Theory of Nature and Mode of Production*, Political Epistemology,
https://doi.org/10.1007/978-3-032-02295-0_1

historians regard not only as a first flowering of the bourgeoisie, but even already of the capitalist mode of production. The northern Italian region not only produced important pioneers of classical natural science and of the bourgeois worldview, above all Leonardo da Vinci (but also Machiavelli, Ficino, and Giordano Bruno); its scientific centres, especially Bologna and Padua, attracted scientists from other European regions (e.g. the Brussels physician Vesalius) and influenced scientific development outside Italy, as is clear from the Nuremberg School and especially Copernicus.

Even more than in this Italian heyday, the fact that scientific development depended on the development of the bourgeois mode of production becomes tangible in its gradual decline in the course of the seventeenth century, when the shift in the centre of world trade from the Mediterranean to the Atlantic made Italy the plaything of the great European feudal powers. German science had a comparable fate, because the development of the German urban bourgeoisie—and this applies to a certain extent even to the North and East German Hanseatic League—was closely linked to the role of the Levant trade. The economic flowering of the German cities in the fifteenth and early sixteenth centuries saw the rise of scholars such as Nicholas of Cusa, those of the Nuremberg School, Agricola, and Paracelsus; and just as Italian early-bourgeois science reached its apogee only during the disintegration of its social preconditions with Galileo, so too did the German with Johannes Kepler. As the European bourgeoisie's geographic centre of power shifted during the course of the sixteenth century from northern Italy and Germany to the Netherlands and England, the centre of the genesis of modern natural science moved at the same time to the Atlantic rim, where this development reached its first conclusion. Seventeenth-century French science, in an attempt to keep up with the times, contributed to this development to an extent corresponding to the constrained possibilities that the French bourgeoisie faced under absolutism—apart from during the reign of Henry IV.

The coincidence in time and place of the development of the bourgeois mode of production and classical natural science sketched here, instructive as it is in my opinion, provides only an *external* connection that invites us to enquire into its *inner* necessity. This inner necessity can perhaps be best initially approached by clarifying to what extent the development of the bourgeois mode of production and thus of the bourgeois class itself was a *precondition* for the development of classical natural science. It is not enough to point out that where the development of the bourgeois mode of production did not progress, nothing decisive happened for the genesis of classical natural science. Rather, it is a question of how exactly the development of the bourgeois mode of production created the preconditions for the development of science.

1.1.2 New Horizon

Let us begin with the material prerequisites, thereby paying no further attention here to the general condition for science that a society's labour productivity permits certain of its members to be temporarily or wholly exempted from material production, because this is not a peculiarity of the development discussed here. The situation was different with regard to the horizon of perception that science needed for its development. For this, the development of trade in goods represented an indispensable prerequisite. The expansion of trade broadened the field of vision, both with regard to nature—the starry sky gained in breadth, the world of minerals, plants and animals multiplied, a "new world" was discovered—and to the human arts, the products of which were brought by merchants from all known regions of the globe. The international trade in goods, which developed into world trade, created, as it were, the material basis for science. Trade opened up a world of objects for science whose horizons extended to infinity, faced with which thinkers were forced to shake off their narrow-mindedness. Also encountering and appropriating the essential scientific results of antiquity, as they had been preserved and developed by the Arabs, who for their part had appropriated the results of the Indian development of science, was to a considerable extent due to the struggle for supremacy of the occidental merchants on the Mediterranean, a struggle whose bourgeois character is hardly concealed by its feudalistic form (crusades), at least from the second half of the twelfth century on. Thus trade made the existing scientific knowledge of the whole world (apart from China) available to occidental science. The development of science under consideration here can thus be shown to depend on the development of the bourgeois mode of production, a form of production for exchange based on the division of labour, which initially developed on the basis of the trade in goods between economic regions (without yet determining the latter's inner structure). It is in this sense that bourgeois production provided science with its objects of study.

The fact that the Italian mathematicians of the fifteenth and sixteenth centuries were able to draw on certain computational techniques which they found developed in mercantile bookkeeping shows that bourgeois development as a precondition of classical natural science can also be examined in terms of the extent to which it even prefigured the procedures required for science.[2]

1.1.3 The Machine as Object of Study

The most important objects of study for the development of classical natural science, however, were supplied only secondarily by trade, but primarily by non-agricultural production, indeed in its means of production. As in antiquity Archimedes developed his mechanics based on the study of the means of production of his time, so also for the expansion of classical physics the study of composite tools and the machines or machine warfare devices of that time was of decisive importance, as the biographies of practically all physicists of this epoch show. H. Grossmann divides these means of production into three groups of objects of

[2] Italian merchants, for example, used the decimal system apart from the Arabs at a time when university mathematics was still struggling with decimal numbers. Cf. Crombie (1952, 34).

study.[3] In the clock and the mill (a collective term for driving and simple working machines),[4] it was possible to study above all the transformation of mechanical movements, the translation of time into space, but above all dynamics, especially the quantitative recording of forces; in lifting devices, practically all statics; finally, the gun stimulated the study of kinematics and made free fall in particular a problem.[5]

Admittedly these complicated tools and machines, which were extraordinarily important for the development of the classical natural sciences, were, for reasons yet to be discussed, by no means typical of the non-agrarian production of the time, but were used on its margins or even—as implements of war—outside of it. Nevertheless, the refinement and higher development of handicraft production, due to the intensified division of labour, was the decisive condition for the development of the advanced means of production, which as an object of study represented an essential prerequisite for the modern natural sciences.

1.1.4 Scientific Instruments

Highly developed artisanal technology, however, not only made these advanced means of production possible; it was also the factual prerequisite of modern science insofar as it gave it the working tools it needed for its progress: printing, telescope, microscope, and all the equipment without which experimental science would have had to remain a mere intention.[6] At the end of the Middle Ages—which in its level of civilization can otherwise hardly be compared with other societies based on the dictatorship of landed property (Antiquity and the so-called Asian modes of production)—a form of artisanal technology emerged, which became a factor in later capitalist development and which interests us here as a precondition of the modern natural science. However, this outstanding result is due to none other than the specific bourgeois form in which artisanal production developed during the Middle Ages. In the hard-won free space of the cities, the artisan class was able to ensure that its development was not chained to the agrarian structure of production

[3] Grossmann (2009, 103) and passim (first 1935, 161).

[4] Cf. on these terms Marx, *Capital* I, MECW 35, 374–5; I thus refer to them as prime movers, which strictly speaking is only transmission, because this difference has no relevance before the eighteenth century in my opinion.

[5] Cf. also B. Hessen (2009). This essay examines in detail the stimuli that classical physics received in all its themes from the problems of the advanced means of production of the time. The tendency of this work, which seems to me to aim at proving that the development of modern natural science corresponded to a need of material production, transport, etc., is something I am admittedly unable to follow, as will be shown in the further course of my argument. The same applies to the Hessen-oriented work of H. Mineur (1970, 39) passim. My critique of Hessen may at first sight seem identical to Clark's critique (1970, Chap. 3). However, the further explanations will make it clear that I do not consider the division of social life into independent spheres such as art, economy, war, etc., on which Clark's critique is based, to be very fruitful.

[6] Cf. D. J. [de Solla] Price (1957, 585 ff. and 624 ff.).

and exploitation dominated by large estates (in contrast to the so-called Asian mode of production); however, the slave labour that had inhibited technical development in antiquity no longer existed.[7] Compared with these favourable factors, the guild-like organisation of craft production, which was soon to prove a serious obstacle to technical development, was at first an underdeveloped moment of the bourgeois form of the development of craft production, and was not yet a major factor. This form itself, however, can be seen here to be a prerequisite for the genesis of classical natural science.

1.1.5 The Social Actors

The development of the bourgeois mode of production must also be regarded as a precondition for the development of classical natural science, inasmuch as without it the social actors who advanced this development could not have come into being.

The evidence that these actors did not as a rule come from the nobility plays no role in this context, which applies equally to the scholars of the Middle Ages. What is more decisive is rather that those who promoted the genesis of modern science were oriented towards the world of appearances; they subjected speculations, as indispensable and popular as they remained, to scrutiny, and were interested above all in the practical result of their actions. This attitude was only able to develop because these early scientific protagonists were familiar with the problems of the technology of their time, whether they themselves had to deal with these problems professionally, or whether they were called upon by the technicians. Thus it is not surprising to see what an enormous proportion of the pioneers of modern science were artisans, engineers, physicians, captains, etc.

1.1.5.1 Vernacular Language and New Academies

More than any biographical survey, this situation is illuminated by the fact that modern science no longer used Latin alone as the language of science, but since the late fifteenth century increasingly used the respective vernacular language. This fact not only reflects the emergence of nation states due to the development of the bourgeoisie, but it also gives expression to a practical necessity. Engineers and artists who were uneducated in the scholastic sense, who did not speak Latin and Greek, or not sufficiently well, played such a decisive role in the development of modern natural science that it could no longer articulate itself solely in the traditional scholarly language.[8]

[7] On the fetters that ancient slavery imposed on the development of natural science, see J. Kuczynski (1972, Chaps. 2 and 3).

[8] 8 Cf. in particular L. Olschki (1919–1927), here vol. 2, 112 passim. Cf. also E.J. Dijksterhuis (1956, 273 ff.).

At the same time, this shows that the traditional educational institutions of science had come into contradiction with the requirements of the latter's development, a contradiction that went so far that the bourgeoisie created its own institutions of education, discussion and research, of which the Accademia del disegno[9] and Gresham College[10] are the most famous examples. It was thus the practitioners of the professions that developed with the bourgeois mode of production, whose problems could no longer be solved using traditional manual routines and thus demanded new solutions, who gave the development of classical natural science its character, at least from the second half of the fifteenth century onwards. Science's character, in turn, also conditioned the activity of professional scientists in the traditional sense, insofar as they did not simply stand apart from this development.[11]

1.1.6 *Was a New Natural Science Needed?*

If all this may convince us that the development of the bourgeois mode of production is to be regarded as a condition of possibility for the genesis of modern natural science, it remains to ask whether, beyond providing these conditions, the bourgeois mode of production also made the development of modern natural science necessary, i.e. whether it added to the possibility the compulsion that it should materialise. Of course, this question cannot be answered with any obvious teleological argumentation that tries to justify the necessity of the development of natural science in the foregoing centuries on the basis of industrialization in the late eighteenth century. Rather, as will become clearer, a characteristic difficulty in attempting to materialistically reconstruct the genesis of modern natural science becomes apparent here. There is no doubt that this natural science is a condition of the later industrial revolution and therefore had to form before it. On the other hand, it is only industrial production that creates the conditions for using science for material production, and thus the need for it. Its genesis cannot, therefore, simply be derived without further ado from the requirements of material production.

To make this clear, it is perhaps useful to imagine as concisely as possible the situation of material production in the sixteenth/seventeenth century with regard to its need for scientific knowledge. In doing so, I will limit myself to examining whether and at which points technical bottlenecks appeared in material production, which it was both economically urgent to overcome but which could no longer be resolved with the conventional methods of material production; in other words, bottlenecks whose resolution imperatively depended upon the natural sciences.

---

[9] Cf. L. Olschki (1919–1927, vol. 2, 171 ff.).

[10] Cf. M. Ornstein (1928, 4).

[11] Cf. Kuczynski (1972, Chap. 4, esp. 97 ff.).

1.2 Second Section

1.2.1 The Situation of Agriculture

No such bottleneck can be observed in the most important sector of material production in the sixteenth and seventeenth centuries, namely *agriculture*. It is true that the development of non-agricultural production placed increased demands on agriculture; it had to produce an increasing amount of raw materials for craft and manufacturing production (especially for textiles), and also of food for a steadily growing urban population. Yet it was able to meet these demands on its existing technical basis. The conversion to wool production (pasture farming) even represented a step backwards in technical terms.

Significant quantitative and qualitative changes in agriculture took place in the sixteenth and seventeenth centuries, however. It was during this period, for instance, that crops introduced from the New World (potato, maize, tomato, etc.) began to make significant changes to the European food supply.[12] Crop rotation was introduced in north-western Europe during this period in addition to the alternation of field and grassland cropping, probably the most significant agronomic innovations since the establishment of three-field farming at the turn of the millennium.[13] But all this only modified the conventional technical basis of agriculture, rather than constituting a new stage of development.[14]

The conventional technical basis of agriculture only produced a bottleneck in the eighteenth century during the transition to the industrial age, when the increased demands of agriculture came to a head and economic development as a whole depended on overcoming this bottleneck. Between 1750 and 1850, geology, physiology and chemistry provided the prerequisites for putting agricultural technology on a scientific footing and thus overcoming the bottleneck that had arisen. The fact that it was these sciences on whose development the scientific grounding of agricultural production depended also demonstrates that the natural science developed in the sixteenth and seventeenth centuries, physics, would not have been able to help if conventional agricultural technology had already constituted a bottleneck at that time.

---

[12] Cf. R.J. Forbes (1957, 1 ff.).

[13] Ibid., p. 12 ff.

[14] W. Jonas, based on the agricultural historian W. Abel, sums up the situation of agriculture in this period as follows (1969, 222): "For the sixteenth and seventeenth centuries, the following development can generally be noted: to 1: means of labour—extensive stagnation. To 2: Cultivation systems—often not insignificant developments. To 3: Man as a productive force—varying in degree, but generally worsening situation […], which sometimes cancelled out the progress made through better farming systems."—On the development of agricultural tools, cf. O. Beaumont and Jenkins (1957, 134 ff.).

1.2.2 The Situation of the Craft Sector

As far as non-agricultural production is concerned, there was no significant bottleneck that needed to be overcome with natural science.

First of all, recall that until the eighteenth century, non-agricultural production was predominantly provided by *handicrafts* or by the work that peasants performed on the side. As much as manufactories and the first factories are to be regarded as the most important form of production of the sixteenth/seventeenth century with regard to the development leading to industry, their contribution to the volume of the total non-agricultural production in this period was still minor.[15] The relationship between handicraft and manufactory or factory production is, quite apart from regional differences, naturally different in the various branches of non-agricultural production. In textile production, in particular, manufactory production had probably already displaced handicraft production from pole position in the seventeenth century.[16] The reason for the unbroken dominance of handicraft production is generally the fact that the development of the market only exceptionally allowed large-scale production that exceeded the dimensions of handicraft—whether manufactory or factory-based. Without a colonial empire which was not only plunderable but also receptive to goods, which was only formed in the eighteenth century, and without the assertion of the supremacy of world trade by the economically most developed nation, which England did not achieve until the eighteenth century, this could not be otherwise. Even in the most developed regions of Europe the formation of commodity production had not yet acquired the breadth and depth that would have made the production of use-value only an insignificant marginal phenomenon alongside it.[17] This was probably the decisive limitation of the market prior to the eighteenth century. With the exception of a few sectors, craft production satisfied the demands made on non-agricultural production under these conditions, especially since it had become extraordinarily refined and developed between the fourteenth and seventeenth centuries.[18]

The expansion of trade during this period led to a significant diversification of craft production[19] and a concomitant deepening of the division of labour (though generally not within the individual craft enterprise), which naturally also led to a further development of craft tools[20]—compare, for example, the spinning wheel or the loom of the thirteenth with that of the seventeenth century.[21]

---

[15] Cf. J. Kulischer (1928–1929, vol. 2, 148 f.).

[16] This is certainly true if one also takes into account the decentralised manufactories, which are admittedly difficult to distinguish from publishing production in individual cases. Cf. H. Mottek (1971–1974). vol. 1, 307 ff; cf. also J. Kulischer (1928–1929), vol. 2, 154 ff.

[17] Cf. E.J. Hobsbawm (1999), M. Dobb (1963), especially Chap. 5.

[18] Cf. F. Klemm (1954, 85 ff), and F.M. Feldhaus (1930/1971, 308 ff. and 333 ff.).

[19] The wealth of textile varieties is particularly impressive, cf. F.J. Flanagan (1957, 187 passim.).

[20] Cf. R.A. Salaman (1957, 134 ff.).

[21] Cf. O. Beaumont and Jenkins (1957, 160 f.).

The technical basis of artisanal production had thus undergone a significant development. This technical development, however, could not in principle advance to a stage at which its further progress would have required a scientific foundation: artisanal production and the use of science in material production were mutually exclusive. One can formulate it more pointedly: where the craft form of production had not itself become the decisive bottleneck in the further development of material production—and, as we have seen, this was not the case at the time—material production did not need science.

1.2.3 Handicraft Production and Science

The reasons for this will be discussed here in somewhat more detail, because they are of fundamental importance for reconstructing the epoch in the history of science under consideration here. For if science and artisanal production (and this is predominant in non-agricultural production) exclude one another, then our claim above that the bourgeois mode of production—even in its artisanal form—is the condition of possibility of the genesis of classical natural science is once again called into question. How can something be a precondition of science that at the same time excludes it?

1.2.3.1 Forms of Application of Natural Forces

The application of scientific knowledge in material production is the intellectual side of a specific form of the use and application of natural forces in material production. All material production is, of course, the application of natural forces; it cannot proceed otherwise. But this application of natural forces takes different forms and correspondingly different forms of its intellectual moment are required for the particular mode of use of natural forces. As long as natural forces are used in work as modes of action and properties of certain natural things, scientific knowledge is not required as the intellectual moment of labour; knowledge of things and their properties, insofar as they are relevant to the work, is all that is needed. Scientific knowledge, on the other hand, is required when it is a question of using natural forces in their general form. This generality is certainly historically relative. Nevertheless, the qualitative difference between these two forms of the use of natural forces can be made clear with the help of a simple example.

It is undoubtedly a use of a force of nature in a non-general form when Darby discovered at the beginning of the eighteenth century that the smelting of iron ore with coke offered great advantages compared with the previous processes;[22] a use of the force of nature in a general form would have consisted in this case in the better

[22] C.f. F. W. Gibbs (1957, 689).

smelting process being discovered based on the recognition of the fact that smelting is essentially a chemical reduction process. If in the first case the reducing power of the process is recognized and utilised only as a property of a certain natural thing, in the second case it is this natural power itself which is recognized and utilised as such.

Of course, the reduction takes place only between particular natural things and does not exist detached from them; the forces of nature in general form—here, chemical forces—do not themselves exist as a natural thing, but only as the lawful form of the effects that natural things exert on one another. Thought comprehends them in their general form by finding out what is identical in the many particular effects between the many particular natural things, by isolating in the natural things the identical moments which are decisive for the force of nature under investigation, and thus by discovering the general conditions under which the natural force acts. This knowledge—together with other conditions—makes a new use of natural forces possible; instead of searching for natural things that exert whatever effect is required in the work process on other natural things in a more or less satisfactory way, it becomes possible, starting from the knowledge of the natural force and its conditions, to compose these conditions using natural things in such a way that the natural force emerges from them and acts in the form needed in each case.

1.2.3.2 Practice and Abstraction

It is obvious that the possibility of such knowledge does not depend primarily on intellectual potential. The discovery of the identical in the many particular processes between natural things and the isolation of the moments in these natural things on which the effect depends—these theoretical abstractions presuppose corresponding abstractions that have taken place in material production. In practice, the distinction between the natural force at stake in each case and the natural things in which it is found can be approached from two different starting points corresponding to those two theoretical abstractions: natural force and natural thing. Practice works out what is identical in the many processes between natural things by including an ever-increasing number of different natural things in its procedures and by becoming capable of provoking the same effect in a multiplicity of different natural things (abstraction by substitution). Practice also isolates the moments of a natural thing that are decisive for a particular effect by working on those natural things, eliminating the non-essential or even obstructive aspects of the natural thing in question and refining the essential ones until they achieve their most effective form (abstraction by isolation). It is easy to see how many millions of times practice must have proceeded in this way, how manifold the materials processed must have been, how developed the processing technique, before thought could succeed in grasping a natural force in its general form.

These preconditions of theoretical abstraction, which were laid down in material production, were intensified in early bourgeois non-agrarian production, which was determined by craftsmanship, in the course of the progressive division of labour and the differentiation of production processes. Material production produced a growing assortment of utilitarian objects using the same materials and, conversely, the same utilitarian objects using a growing number of different materials; it required tools and substances to form an ever broader range of materials and, conversely, an ever more differentiated arsenal of tools and materials to form the same products. In this process, material production revealed the multifaceted properties and modes of operation of the tools and objects used for work and created possibilities for comparing and gradually conceiving material processes in their general form.

1.2.3.3 The Guild Structure

At the same time, however, craft production prevented these possibilities from being realised. The guild-like organisation of the crafts plays only a marginal role here, although it undoubtedly represented a formidable obstacle to any innovation.[23] (The still powerful fetters of the guild system in the sixteenth/seventeenth century, even in the most developed regions of Europe, the poor contribution made by the cities of Brabant to the Dutch Revolution, and the compulsion to locate manufactories outside the cities in England even in the early seventeenth century, provide sufficient testimony to this fact).[24] Even more than the barriers of the guild system, the craftsman's own stunted economic structure prevented him from envisaging innovations towards large-scale production; he lacked not only the necessary overview of the market beyond local borders, but above all the necessary capital. This is probably the main reason why the merchants also took the initiative in the emergence of manufactories, as in the establishment of the putting-out system, and factories took advantage of the available opportunities for large-scale production.[25] But these considerations are of only secondary importance in our context; they become significant only under the following conditions. First of all, we must trace how the form of artisanal production determines and limits its intellectual aspect.

1.2.3.4 Personal Unity of Theory and Practice

Artisanal production is the business of an individual producer, who is a specialist within the framework of the social division of labour, while as a rule he is a comprehensive producer with regard to the product of his trade. The commercial side of his business is also in his own hands. Just as he masters all the work necessary to produce his product, so he must possess all the knowledge necessary for this purpose.

---

[23] F.M. Feldhaus (1930/1971, 338 f.), gives an instructive example.

[24] With regard to the establishment of manufactories and factories in the countryside, a distinction must be made in each case as to whether this choice of location is due to the power of the guilds in the cities or to the use of water mills as drives or simple working machines.

[25] This is true even for the seventeenth century French manufactories, which were artificially created by the absolutist state.

The individual producer owns the means of labour he requires, and unites in his person the physical and intellectual faculties required to manufacture his product. The individual producer thus represents, as it were, the unmediated, personal unity of theory and practice.

We must not conclude, however, on the basis of the fact that he possesses both the physical and intellectual faculties necessary for his work that the individual producer's knowledge is of a purely individual character. On the contrary, this knowledge is general in two ways: on the one hand, it is the knowledge of a profession; that is to say, the individual producer has received his knowledge from the professional peers of the preceding generation and shares it—within certain limits (comparable level of technical development, comparable working materials, etc.)—with the professional peers of his time, whereby communication among these peers increases with the expansion of trading relations. On the other hand, the knowledge of the artisan producer is general in the simple sense that it necessarily reflects the moments of the work process objectively correctly, however superficial this reflection may appear from a more advanced stage of knowledge.

1.2.3.5 'Natural' Division of Labour

Admittedly, this knowledge is necessarily limited in several senses. First of all, this knowledge and its scope are indissolubly linked to and determined by the practical production of a particular product. The artisanal producer can only trace the interrelationships between his means and objects of work within the narrow limits imposed by the compulsion to produce a saleable product. As a rule, he cannot afford to experiment beyond these limits. The potentialities of his working materials thus come into his view only one-sidedly, in specific reference to the product to be manufactured. This one-sidedness is cemented by the social division of labour, a division of labour which, given the limited mobility of property and thus of people in the early stage of development of the bourgeoisie, dictated not only an individual's activity but that of entire generations. The artisanal producer was at best superficially aware of the versatile use of the means and objects of labour that emerged in the social context precisely because of the ever-deepening division of labour.

This means that it was impossible for the artisanal producer to see the possibilities that gradually emerged in the processes performed in material production in generalised form due to the deepening of the societal division of labour and differentiation of production. He had no time, no opportunity, and no means to perceive them. The intellectual appropriation of that generalisation carried out by material practice presupposed the abolition of the immediate, personal unity of theory and practice characteristic of artisanal production; for this immediate unity itself fell under the "natural" [*naturwüchsig*] societal division of labour and thus remained particular. In relation to this particular unity of theory and practice, under these conditions, a division of labour concerning the relation of theory and practice itself must be regarded as the step that made it possible to summarise and assess the insights gained in the labour process. The generalisation of the knowledge acquired in manual production demands an intellectual activity which is to a certain extent removed from the "natural" division of labour in material production.[26] (Technical literature, which began to appear in the fifteenth century, represents the beginning of a fundamental change in the form of the intellectual moment of material production in its attempt not only to collect the wealth of experience acquired by craftsmen, but also to systematise it and to assign to it pieces of scientific theory which seem appropriate.[27] The authors—Agricola, Alberti, Biringuccio, Filarete, Ramelli, Vives, etc.—have, of course, little in common with master craftsmen. Their writings form the counterpart of the profession of the engineer, which was developing at this time, emerging in production sectors which transcended the limits of the crafts, and in which the unity of intellectual and manual labour, peculiar to the craftsman, was abolished. These engineers, however, are by no means scientists, and the borrowings they make from natural science are not—as the third chapter of this book will show—to be regarded as an application of science in material production).[28]

The contradictory significance of the division of labour for the theoretical comprehension of the labour process also sheds light on the fact that craft production within a trade did not, as a rule, know any division of labour. Thus the differentiation of the individual moments of the labour process was limited, and these moments could not be developed to their fullest and at the same time simplest form. As the importance of organic manufacturing[29] for constructing working machines would later show, the theoretical comprehension of the labour process essential for this depended on the moments of the process being practically distinguished and shaped into the form that was at once most effective and simplest. This comprehension was closed to handicraft production, which had no possibility and no necessity for it. If the social division of labour condemned the artisan producer to a narrow specialism, the lack of division of labour in his enterprise meant that his specialised knowledge could not attain real depth.

1.2.3.6 The Decisive Productive Force

All these limitations are grounded in the individual form of craft production; but this in turn is both founded in and at the same time determines the state of the forces of production. The manual dexterity of the individual is the most developed and—if we disregard the biological natural production used in agriculture—most decisive force of production of the early bourgeois epoch. At this time, the transformation of natural things into objects of use was predominantly dependent on man's physical capacity for work and the development and training of this capacity; hence the

[26] Cf. V. V. Bykov (1974, 78 passim.).

[27] Cf. besides L. Olschki, op. cit., F.M. Feldhaus, who documents illuminated manuscripts of the fifteenth century (1930/1971, 339 ff., 376 ff., 381 ff.).

[28] Cf. Olschki (1919–1927, esp. vol. 1, 30–44); cf. also G. Harig (1958, 419–450); W. B. Parsons (1939), Zilsel (1942) and Zilsel (1926, 2. chapter, esp. 130–158.).

[29] Cf. on this term K. Marx, *Capital* I.

actual process of transformation stands and falls with the potentials contained in man's bodily organs of labour. While coarse work (moving heavy objects, etc.) could be performed cooperatively, if natural forces were not already used in the process, the individual producer was responsible for precisely the normal and even more so the finer and more complicated processes of transformation. As the manufactories show, individual producers could only cooperate on the basis of parallel activity, or by dissecting the labour process into partial steps that could be accomplished in individual production. In other words, individual labour could not be abolished by cooperation between individual producers, although, as will be shown, this cooperation had important consequences for the intellectual moment of labour. As later industry shows, individual labour could only be fundamentally overcome with collective means of labour.[30] One does not encounter such means of labour in the actual transformation process—apart from in certain rough work—until the middle of the eighteenth century. The tools of craft production are not only tailored to the physique of an individual (the average strength of a man must be sufficient to handle them, they must not require hands with two thumbs, etc.); they are also nothing more than the extension of the bodily organs of an individual. It is not the carving knife that shapes the wood, but the hand that guides it. The bodily organs of labour are the decisive means of labour, the tools only accessories of these organs. It is true that the tools mediate the effects of the bodily organs of labour on the objects of labour and thereby determine the evolution of these organs; but in this they remain determined by the natural constitution of these organs, which can be modified only within limits. The development of the dexterity of the bodily organs of labour depended on a further division of labour more than on tools; this is demonstrated by the manufactories of the eighteenth century, which exceeded the high level of development that the art of craftsmanship had achieved in the sixteenth and seventeenth centuries through an extremely intensive division of labour.

1.2.3.7 The Barrier of Generalisation

The formability of the dexterity and capacity of the bodily organs of labour, limited by their natural constitution, thus constitutes the principal barrier at this stage of the development of production. The fact that at that time the physical working capacity of the individual was the decisive "natural force" applied in non-agricultural production also implies the incompatibility in principle between craft production and the use of scientific knowledge for material production.[31] For this means that the

[30] Cf. A.A. Kusin (1970, 39).

[31] In the non-agrarian production of that time, there was only one mode of formation that was not determined by the physicality of the individual: the chemical one. It is true that the chemical process of formation is set in motion by the bodily organs and their tools, but it proceeds independently of them. Now, on the other hand, it is clear that the formation of a scientific chemistry had to overcome much more difficult obstacles than, for example, physics. Bernal (1971, vol. II, 475) remarks: "Chemistry, unlike physics, demands a multiplicity of experience and does not contain

mechanical natural force applied in the actual process of transformation of natural things into objects of labour was only available as a property of a certain "natural thing," the physis of man, i.e. not in a general form. For a long time this was true even of the simple (non-specific) exertion of mechanical force; later men made use of the bodily forces of tamed animals for this purpose. However, the scientific formulation of the general concept of mechanical force in classical physics (Baliani and Newton) presupposed the use of further and less complicated natural things, the use of wind and waterpower. Until the industrial age, however, the physical capacity of man to work remained indispensable for all more specific mechanical transformation. Mechanical forces of nature could not be applied in a general form, and thus science could not be either.

It is not accurate enough to say, however, that for these reasons scientific knowledge had no significance for the core area of non-agricultural production in the sixteenth/seventeenth centuries. It would be more precise to say that, under these circumstances, it only had a very specific significance in this domain, namely contributing to the development of the possibility of using natural forces in a general form for specific processes of transformation.

As has already been argued in the preceding section, there was no economic necessity for this before the eighteenth century. Moreover, it is obvious that science would by no means have had the main role to play in this development. In this case, too, a contribution of the natural sciences presupposes pre-scientific solutions provided by practice. Machines in which the use of natural forces is laid out in general form and on which these forces can be studied are the prerequisite and not the result of the use of scientific knowledge for material production. This is demonstrated by the history of mechanical engineering up to the nineteenth century. It was not scientists but engineer-craftsmen who designed and built machines. However, from the late seventeenth century onwards, ideas developed in the natural sciences played an increasingly important role in the design solutions. It is probably from this historical point onwards that science no longer owes everything to practice, but practice also owes some things to science. Until the invention of the mechanical loom and the spinning machine in the eighteenth century, machine development did not concern the core area of non-agricultural production, and all the prerequisites for using scientific knowledge for material production were therefore lacking in this area.

self-evident principles." It can remain undiscussed here whether the sequence in the formation of scientific physics and scientific chemistry was not also conditioned by the fact that chemistry presupposed physical atomism, as little as chemistry could directly begin with it, for its breakthrough to science.

1.2.4 The Situation of the Manufactories

Before turning to the general consequences that this inapplicability of scientific knowledge, due to the state of the forces of production, has for the vast majority of material production, we must determine whether and where bottlenecks occurred in the material production of the sixteenth/seventeenth century, which made the application of scientific findings in material production necessary.

As is already apparent from the foregoing considerations, the same negative answer as for the craft trade also applies to the *manufactories*, since the human being's organs of labour represent the frame of possibilities and thus the limit of manufactory, just as they do for craft production. The dissection of the work process, which the craftsman used to perform from start to finish, as carried out in some manufactories, brought with it the need to coordinate the partial aspects, as well as the dissolution of the immediate unity of theory and practice characteristic of the craftsman. The analysis and modelling of the work process become work steps detached from the completion of the work as a whole. Thus, as in the dissection of the formerly holistic work process, important conditions undoubtedly arise that contribute to the later development of machines that actually transform natural things into objects of labour.[32]

Thus if in manufactory production we can undoubtedly identify steps that are extremely important for the subsequent application of scientific knowledge in the core area of material production, these are by no means however the bottlenecks we were looking for. We must therefore ultimately examine the more marginal sectors of non-agricultural production that are promising with regard to our question, because of the mechanical and simple industrial machinery they employ.

[32] Cf. S. W. Schuchardin (1963, 76). Since here a superficial glance might discover a proximity to Borkenau's "Manufaktur" section (F. Borkenau 1934, 1–14), I will not only point out that I share Grossmann's criticism of Borkenau (cf. H. Grossmann 2009), but also note the following: Borkenau attempts to show in the result of organic manufacture, i.e. in the partial labour reduced to certain manual operations, the economic labour category that Marx found out for bourgeois society, to a certain extent technically realised. In doing so, I do not think he misses this category alone, whose generality is not based on the reduction of labour to a simple basic labour, but on the contrary on the universal diversity of labour from a certain stage of development of bourgeois society (cf. K. Marx *Outlines* (*Grundrisse*), MECW 29, Introduction). If one were to follow Borkenau, it would not be plausible that the division of labour through organic manufacture could be regarded as a prerequisite for the development of machines for the actual processes of formation, since it is precisely a matter of finding machine solutions for the specific processes of formation; machines had long since been developed for Borkenaa's concept of labour, which was reduced to purely mechanical expressions of force.

1.2.5 A General Technological Bottleneck

In the seventeenth century, a bottleneck in the sense we are looking for began to emerge, above all in one sector of non-agricultural production, namely *mining,* which revealed the limits of the entire technical basis of this stage of development. The advance of mining into greater depths depended decisively on water drainage, but this had a weak point in the pumps, or more precisely: in the drives of the pumps.[33] The drainage pumps in the mining industry proved the state of use of natural driving forces achieved in the seventeenth century to be insufficient.

In addition to the use of animal power, the use of wind and waterpower by mills had already occurred in the Middle Ages. Building on the mill technology of the Middle Ages, the use of waterpower, which was superior to animal power in terms of cheapness and strength, and to wind power in terms of continuity, was considerably further developed and intensified from the fifteenth to the seventeenth century in accordance with the requirements of the growing use of machines, and especially of mining. This development took place on a craft and thus pre-scientific basis, even if the problems that arose in the process, e.g. the quantitative determination of forces, increasingly made the insufficiency of this craft basis perceptible. The impetus that science received from these problems cannot be overestimated, even if science at that time could not help to solve any of the problems that arose.

In the seventeenth century, the use of waterpower on an artisanal basis had thus reached the point where it was not the inadequacy of this artisanal basis but the inadequacy of the waterpower itself used by mills that began to emerge. The water-lifting machine at Marly (Seine) may serve as a somewhat curious example of this emerging frontier.[34] This enormous machine, built in 1681–1685, was intended to raise the water used for the fountains in the park of Versailles from the Seine to the required height. This machine reached the then sensational power of 80 hp and was so expensive that even Louis XIV could not afford to buy it a second time. The deployment of such cutting-edge technology in material production was therefore ruled out for economic reasons. But how could water be drained from deeper mines if the 80 hp of Marly could not even be considered to drive the pumps? It was not by chance that the development of the steam engine in the first half of the eighteenth century began with the problem of driving mine pumps. The history of the steam engine shows the circular and comprehensive nature of this bottleneck.[35]

The idea of using steam power instead of waterpower dates back to the first half of the seventeenth century. The analogy with waterpower pointed engineers in the direction of the steam turbine, as designed by G. Branca in 1629. However, the high speed of the turbine required a material for the paddle wheel and especially for the gearbox with properties that no known material had (it was not until the second half of the nineteenth century, a quarter of a century after the development of modern

[33] Cf. F. Klemm, op. cit., p. 207 ff.

[34] Cf. ibid., p. 204.

[35] Cf. ibid., p. 219 ff.

steel production, that the steam turbine became feasible: it was patented in 1889 by Laval). The revolution in materials that became necessary, however, depended on smelting technology and—after the excellent suitability of coke for smelting iron ore had been discovered in the eighteenth century—on the development of mining.[36] The mining industry, however, was only able to meet the growing demands of coal by revolutionising the drive of the pumps. The circle closes, and at the same time it becomes apparent that this is not only a weak point for mining, but for the further development of the entire technology of the time. Against this background, the genius of Watt's piston steam engine can be seen in the fact that—in contrast to the steam turbine—it was the "detour" that harmonised with all the factors of the technical stage that had been reached in order to take the decisive step forward in the use of natural forces for propulsion.[37]

Although Watt, as well as the inventors of the most important precursors of his machine, Savery and Newcomen, were not scientists but engineers, it is nevertheless justified to see in the steam engine one of the first cases of conscious application of ideas which have their origin in natural science (latent heat, vacuum).[38] However, as much as e.g. the vacuum problem was researched in the seventeenth century in connection with problems to do with the pump (Boyle, Guericke, Pascal, Toricelli), it would be difficult to claim that Watt and the other engineers were able to appeal to scientific knowledge that had been developed with regard to that emerging bottleneck. Especially in the case of the vacuum, the expansion of atomism against scholastic natural philosophy was certainly in the foreground for the scientists.

1.2.6 Sectors of Integrated Production

A comparable bottleneck cannot be identified in the *building sector*, nor in the field of *shipbuilding*.[39] These sectors, as well as the development of the mill (clock) and its gear, are on the contrary outstanding evidence of the efficiency of production on a craft basis, which was initially quite able to meet the considerably increased demands of the early bourgeois era. In addition to the mill, the lifting devices[40] of the building trade and the ports[41] are probably the most important objects of study that craft production provided for the development of the natural sciences. What distinguishes these sectors, just like mining and smelting, even more than manufactories, is the integration of various work processes and professions into a total work process, which gives rise to the type of engineer virtually unknown to the Middle Ages and indispensable for the development of the modern natural sciences. The

---

[36] Cf. C. St. Smith et al. (1957) u. R.J. Forbes (1957, 78 ff.), esp. S. 80; cf. also F. W. Gibbs (1957, 685 f.).

[37] Cf. W. Jonas (1969, 230).

[38] Cf. Bernal op. cit., Vol. II, p. 472 f. and 491.

[39] Cf. F.M. Feldhaus (1930, 389), and G.P.B. Naish (1957, 486–93).

[40] Cf. F.M. Feldhaus (1930, 327 f.).

[41] Cf. M. R. Daumas (1969, 399 ff.).

intellectual projection and direction of such integrative large-scale production was incumbent upon these figures, although they cannot be regarded as scientists and their use of scientific theory cannot be seen as simply applying science to material production (cf. above '"Natural" division of labour' onwards).[42] Here, the first decisive dissolution of the craft form of *organisation* of production took place, which sundered the immediate unity of theory and practice peculiar to it and made it necessary to collect and systematise the knowledge scattered in the various crafts and to make the first borrowings from science. The technical basis of these sectors, of course, remained the craftsman's; for here, too, natural forces could be employed by means of machines only for the coarsest work. The *war technology* of the time, which in any case partly overlapped with these sectors, should be classified here, since the provisions just made also apply to it without reservation.[43]

1.2.7 Navigation and Astronomy

To conclude this overview of material production, some remarks on navigation, i.e. on the importance of scientific knowledge for the transportation of the time, are necessary.[44] In fact, I think this is the only case that allows us to speak of the application of scientific knowledge to material production in the early bourgeois era (the calculation of time based on the same astronomical principles is of course also intended here). Certainly, this is an application of scientific knowledge inherited from antiquity, but by no means one that became possible only with the upheaval of astronomy in the sixteenth/seventeenth century, which was so decisive for the genesis of the classical natural sciences. It is even disputed whether the Copernican upheaval of astronomy—whose consequences for the view of nature are not yet of interest here—represented a gain for the navigation of the time. The dispute about whether calculation according to the Copernican system had really become easier compared to the Ptolemaic one cannot be decided by a non-specialist. That at least the results could not at first have been much better than those delivered by Ptolemaic astronomy is already evident from the fact that Copernicus used the observational data of the Ptolemies unchecked. Thus, the use of the "Prussian Tables" (1551), which were calculated on the basis of the Copernican system, was probably more based on ideological sympathy than on their actual superiority over the Alfonsian Tables.

Indirectly, however, the Copernican revolution led to an improvement in the foundations of navigation. The ideological dispute about the heliocentric system led to an intensified observation of the stars and their orbits and to a decisive improvement of the empirical database. Here, above all, lies the merit of Tycho Brahe, who was willing to compromise on the system question, and without whose observational work Kepler's, Galileo's and finally Newton's deciphering of celestial

---

[42] Cf. especially W.B. Parsons (1939).

[43] Cf. F.M. Feldhaus (1930, 325 ff.) and 361 ff; cf. also A.R. Hall (1957, 360 ff. and 369 ff.).

[44] Cf. E.G.R. Taylor (1957, 544–53.).

mechanics would be inconceivable. This improvement in observational accuracy came in handy for maritime shipping, as the inaccuracies tolerable on the Baltic and Mediterranean were a hindrance for Atlantic seafaring.

The problem of determining longitude, which came to the fore with Atlantic shipping, was particularly dependent on the improvement of observational accuracy, as far as the position of the moon in relation to the constellations was used for this purpose—but not on the heliocentric system.[45] The solution to the problem that satisfied navigational practice consisted in developing an accurate transportable clock, which was finally created in the seventeenth century in cooperation between scientists and practitioners (Huygens and Hook).[46]

But even if the astronomical system question was not of decisive importance for navigational practice, the assistance that astronomical science—and astronomy is undoubtedly science even in its Ptolemaic form—provided for Atlantic navigation demonstrated the possibilities of using scientific knowledge for practical purposes. In astronomy, science foreshadowed its future contribution to material production, a promise that inspired the pursuit of science, even if the science of the time could hardly fulfil this promise.

1.3 Third Section

1.3.1 Taking Stock

First it should be noted that before the eighteenth century, the vast majority, indeed the core, of material production did not motivate the development of classical natural science in the sense that there would have been a need for scientific knowledge to solve practical problems. If agriculture, which was the most important production sector, could in any case perhaps not have been expected to provide such an impetus, it happened that the economic conditions and thus the necessity to overturn artisanal production were lacking until the eighteenth century, also for the essential part of non-agricultural production. This fact is therefore extraordinarily important for reconstructing the genesis of modern natural science, because production on a handicraft basis and application of scientific knowledge in material production are mutually exclusive, whereby two moments had to be distinguished. On the one hand, the immediate, the unity of theory and practice in the person of the artisan made it impossible, due to its necessary limitations, to push the intellectual conditions of work beyond the form of particular, casuistic knowledge of artisanal production. On the other hand, however, the underlying state of development of forces of handicraft production excludes the use of natural forces in a general form, and thus the application of scientific knowledge in the process of transforming natural

[45] Ibid, 551 f.

[46] Cf. H. A. Lloyd (1957, 663.).

things. Even for the exceptional sectors of material production in which the artisanal form of production was transcended, science at best provided the explanation for solutions that practitioners discovered without it, even though ideas developed in the scientific community may have provided suggestions here and there. Prior to the eighteenth century, pre-scientific technology was still able to cope with the problems of practice, and the natural sciences were at that time generally not yet in a position to contribute to solving practical problems.

This, if you will, negative result requires us not only to identify the driving force behind the development of the modern natural sciences in something that is at most connected in a mediated way with the needs of material production, but not directly in these needs themselves. Since at the beginning of this study, the hypothesis was advanced that the bourgeois mode of production was the prerequisite for the development of natural science, this result also requires us to ask more precisely to what extent this was the case given the development of the bourgeois mode of production by the seventeenth century. For after all, we must ask how natural science, in the face of a material production that excludes science as an intellectual moment of labour, could develop in such a way that it could later be applied in material production. Neither in material production nor because of its immediate needs could this natural science develop for the reasons discussed; because of the stage of development of material production, there was a necessary gulf between it and science. Here seems to me to lie one of the main problems with which the history of science is confronted when reconstructing the development of the natural sciences in the epoch before the industrial age.

1.3.2 *Mediatedness of the Relationship to Nature and Theory of Nature*

Our conclusion does not merely relate to the fact that the genesis of modern natural science does not arise from the immediate needs of material production. Rather, it was precisely the immediate relation to nature characteristic of the artisanal form of production that excluded the application of science in material production and thus precluded any need for science. With regard to this problem, the view that the behaviour of human beings towards nature, mediated by the circulation of commodities, is to be seen as overcoming the immediate form of behaviour towards nature, which excludes theory, and thus as a decisive prerequisite for the genesis of modern natural science (and also of the natural philosophy of antiquity), is of particular interest.[47]

[47] This view can be found in different variants, which are affected differently by the later critical remarks. The most important representatives—at least in the current reception here—are perhaps: A. Sohn-Rethel (1978), and P. Ruben (1969).

The first thing that seems important to me about this view is that it emphasises the necessary mediatedness of a theoretical attitude towards nature. If this is not supposed to refer to any old theory of nature, but to that which later, from the Industrial Revolution onwards, was applicable in material production, and which could therefore only develop based on material production, as little as material production needed it, then the mediatedness of the intellectual moment of labour acquires special importance for the development of this view.

Now it is obvious that scientific knowledge is not distinguished from the knowledge of a craftsman by the fact that it is mediated at all, whereas the knowledge of the craftsman would be immediate. The *intellectual* moment of labour is of course always mediated, generally through the work itself, but in particular by the means of labour and the social form of labour. In our context we are concerned with the specific form of mediatedness of the intellectual moment of labour and its significance for the development of a theoretical attitude towards nature. The two moments that were distinguished when determining the incompatibility of artisanal production with the application of scientific knowledge in material production represent undeveloped forms of mediation of the intellectual moment of labour: on the one hand, mediation undeveloped by the means of labour (tools) and on the other hand another mediation undeveloped by the social form of labour.

1.3.2.1 Mediation by Means of Labour

As for the means of labour, it appeared that in handicraft production these do not significantly mediate the practical behaviour of man towards nature, and thus neither do they mediate his intellectual behaviour. It is true that the characteristic means of labour necessary for artisanal production, the tool, mediates the practical behaviour of man towards nature insofar as it "guides", as Marx put it, the activity of the bodily organs of labour to the objects of labour.[48] But the nature of the tool is not such that it could form nature by itself; it is the body's working organs that form nature with the help of tools, and thus the practical behaviour of man to nature is an immediate one. The intellectual behaviour towards nature as tried and tested in practice is based on a relationship determined by the physical constitution of the human being; nature reveals itself to the intellect only in the medium of a limited, special relationship. Breaking out of this framework, which is necessary for a theoretical attitude to nature, depends on means of labour that form nature in place of the bodily organs of labour. In other words, scientific knowledge as a form of the intellectual moment of labour presuppose, as its mediator, means of labour capable of shaping nature by themselves.

As we have seen, such means of labour, which should be seen as condition of mediatedness of the intellectual moment of labour, itself in turn the condition of possibility for the genesis of modern natural science, were created in the period

[48] K. Marx, *Capital* I, MECW 35, 190–1.

under consideration here, even if they were not created in the core area of material production. This demonstrates the inner reason why studying machines and their problems was of crucial importance for the development of the classical natural sciences, even if scientists could not initially contribute directly to solving these problems.

1.3.2.2 Mediation Through the Social Form of Labour

The undeveloped mediation of practical and thus also *intellectual* behaviour toward nature is even more obvious when we consider the social form of labour on a craft basis, i.e. individual labour. This individual labour is, of course, the result of a highly developed state of the forces of production compared with so-called prehistory, and in this respect is historically mediated. Moreover, the social division of labour, which makes the individual producer an expert in a special sector, shows him to be dependent on the labour of others, and his practical conduct towards nature to be mediated by that of other members of society. But the result of these mediations is labour in his particular field of expertise, the performance of which is not mediated, or at least not essentially mediated, by the work of other producers. The limited form of the intellectual moment of labour that follows from this mediation of practical behaviour towards nature undeveloped by the social form of labour has been discussed above in its contradictory character for scientific knowledge. It can be characterised by the term "experience," which refers not only to the casuistic of this form of the intellectual moment of labour, but equally to the mode of communication and transmission of expertise corresponding to it, for which mimetic learning, the acquisition of skills and competencies, and not least secrets are characteristic (cf. the section 'Conditions of contradiction' below and above in the section on the '"Natural" division of labour'). It was also shown that this limitation of the intellectual moment of labour, characterised as an immediate, unity of the intellectual and material conditions of labour in the person of the artisan, underwent its dissolution in certain manufactories, shipyards, factories, in certain large-scale buildings, etc., on account of the social form of labour that was being developed there.

This social form of labour differs from cooperation in the early days of mankind in that it builds on the developed level that the various cooperating activities have reached at the stage of a specialised division of labour. It is therefore not based on the principle of interchangeability of the roles of the co-operators due to their lack of specialisation, which would allow the overall process to be coordinated and its sub-moments to be carried out by each of the participants at the same time. In other words, it would be possible to fall back on the unity of intellectual and material working conditions in the figure of a single person. Rather, this form of social labour, based on the developed specificity of its individual moments, demands a change in the form of the intellectual moment of labour.

The intellectual conditions of labour can no longer be those of one person, but must rather be those of a functionally differentiated collective worker

[*Gesamtarbeiter*]. Just as each part of the labour process is mediated by the other parts, so the individual partial process requires knowledge that takes into account the function of the partial process in the total process. The proof of this knowledge can only be experienced in the interaction between the various partial processes. The reflexive structure that emerges as the intellectual form of the moment of labour appropriate to this social form of labour differs from that which was peculiar to the artisan producer in that the mutually mediating moments of labour have been stripped of the appearance of a natural unity given in the person of the artisan, and must first be brought into unity via a mediating activity. The mediation of the particular activities into a total process becomes its own moment of the work process, whose peculiarity consists in ensuring the general side of the process, the interaction of its moments.[49] If the intellectual moment of labour has thus overcome the stage of comprehending the labour process as a natural process, as it were determined by the relation of the individual producer to nature, it thus steps out of this special relation by dealing with the relations between the parts of the work process and thereby comprehending the individual moments under the universal conditions of the total process, i.e. in a general way. Ultimately, the intellectual moment of labour then also acquires the appearance appropriate to it of a creative factor. Its generality remains limited, of course, if the partial moments of the total process remain determined as manual labour by the particular natural powers of the human physique, and thus also their interaction. This development of science would be inconceivable without the emergence of engineers in the context of such social production, a fact that proves that the developed form of the mediatedness of the intellectual moment of labour by its social form is a condition for the genesis of modern natural science.

Despite the fact that scientific knowledge was inapplicable to the material production of the time, in some areas of material production of the early bourgeois epoch, a form of mediation of the intellectual moment of labour nevertheless developed that makes it possible to explain the emergence of a theory of nature oriented toward material production that later, from industrialisation on, became a factor of production. This developed form of mediation was based on the use of machines and on the social form labour that emerged in the first forms of capitalist production.

This two-sided form—on the one hand the means of labour and on the other the social form of labour—of a developed mediation of the intellectual moment of labour in material production itself constitutes the specific condition for the theory of nature to overcome the status of speculation and to become natural science. The limitation of the intellectual moment of labour to pre-scientific, particular knowledge was based on an undeveloped form of labour. As long as this undeveloped form of labour was not overcome, all theories of nature could only be speculative and remote from practice, which itself could have no relevance whatsoever for

[49] This applies regardless of whether this mediating moment of labour is carried out as an exclusive or delegated matter of certain individuals or as a separate step of labour by the collective; whatever the form of implementation—the specialisation of this moment of labour results from this social labour.

production. Ancient natural philosophy, in spite of, or precisely because of, its admirable ingenuity, is the best example of such bottomless speculation remote from practice. The intellectual moment of labour mediated by the means of labour, the machine, and by social labour does not, of course, immediately have a scientific form; the contradictions between this intellectual moment and science will occupy us later (in the third chapter). But it does constitute the specific condition for the modern theory of nature to be able to orient itself at all in its development towards material production.

1.3.2.3 Mediation Through the Exchange of Goods?

Here, too, the weakness of the view that relies on the mediation of the relationship to nature through commodity exchange comes to light, since it does not explain the specificity of the development of science in the modern era. The development of modern natural science is not about the genesis of a theoretical attitude towards nature in general, but about the question of how such a theoretical attitude towards nature could develop, which was able to become the form of the intellectual moment of labour when all the conditions for it had matured. It is difficult to see how such a theoretical relation to nature could be grounded in the relation to the conditions and results of production as exchange values, since in this relation it is precisely the natural nature of these things that is dispensed with, and their relations as commodities say nothing about their relations on the basis of their physical properties.[50] With regard to natural science, abstracting from concrete production inherent in the act of exchange could, on the contrary, only establish a theory which, in its speculative character, would have as little to do with modern natural science as scholastic natural philosophy.[51]

[50] The abstraction of nature as a condition of natural science will probably only be able to have a place in a philosophical concept that regards science in principle as an autonomous positing of the mind.

[51] With regard to Sohn-Rethel's attempt (1978) to prove that the forms of abstraction of the exchange act are the basic pattern of the abstractions of classical physics, Marx's *bon mot* directed against Proudhon should probably be modified to the effect that Sohn-Rethel can at best make an impression on natural scientists as economists and on economists as natural scientists. Incidentally, I see no reason to recognise real abstractions as the foundation of scientific knowledge solely or predominantly in the act of exchange, but not equally in the substitutions and isolations of material labour discussed. To what extent the commodity form was the general form of labour conditions and results in the period under consideration here, and insofar a truly general form, is to be examined separately in the second chapter.

1.3.3 Theory of Nature and the Relation of Production

This critique is in no way intended to deny the importance of social relations as a whole for the development of science, nor even to advocate a one-sided dependence of the natural sciences on the forces of production. As important as the state of development of material production is for the development of the natural sciences, this state of development itself is of course dependent on the relations of production (of which circulation is of course only one moment). This can be seen in the two moments that help the intellectual moment of labour to become a developed form of mediation: in the machine and in social labour. Both presuppose not only a fairly high level of commodity production and of national and international commodity exchange. Social labour in the form discussed arises with the utilisation of capital in production; it therefore presupposes not only a developed market, but producers separated from the ownership of the means of production and a concentration of these means of production and of life [*Produktions- und Lebensmittel*] on the part of the capitalist. It seems important, however, to see in this not only the separation of producers from the means of production and their concentration in the hands of the capitalist, but also that in this way the limited private production of the individual producer is abolished and planned social production comes into being, at least on the factory level. Even if this develops as a higher form of private production and on the basis of the contradictions of a society determined by private production, nevertheless the fundamental pole of contradiction to private production itself develops within it. From here it only becomes properly clear that the dissolution of private labour is the decisive condition for the development of general or "universal labour," as Marx characterised science.[52] The machine, however, is the means of labour that lives up to this dissolution. As far as its social conditions are concerned, it will not be easy to decide in individual cases for the period under consideration here whether the use of machines was already calculated to extract more unpaid labour, whether it reflected a respective lack of labour forces, or whether overcoming the limitations of the human physique, which were especially prominent in rough work, was in the foreground.[53] Perhaps more important than the conditions of the use of the machine are the social conditions of its development. The state of craft technology required for this—which is characterised by a profound division of labour, a variety of products and materials, and a correspondingly differentiated arsenal of specialised tools and materials—would probably have been inconceivable without the specific bourgeois form in which Western craft developed, in contrast to that in antiquity and Asian societies.

---

[52] Cf. K. Marx, *Capital* III, MECW 37, 106.

[53] Cf. H. Grossmann (2009, 133 (1935, 199 f.)). Grossmann cites Cardano's 'De subtilitate', in which he distinguishes four advantages of the use of machines: "1. the saving of human labour, 2. the possibility of using unskilled, i.e. cheaper workers, 3. the reduction of waste, … 4. general advantages of hygiene, which eliminates cleaning costs."

1.3.4 An Unanswered Question

Moreover, it seems to me unquestionable that especially in an epoch in which natural science could not yet be a direct productive force, the generalisation of natural knowledge into an image of nature and the world [*Natur- und Weltbild*], which has a retroactive effect on all individual findings, should be decisively determined by social conditions, i.e. in this case by the conditions of a commodity-producing society. But if we are to explain why the view of nature determined by these social conditions made possible and promoted the development of knowledge of nature into a natural science that proves itself in material production, then these social conditions must be conducive to the theoretical generalisations being based on the conditions of generalisation that develop in material production. Investigating how the social conditions of the early bourgeois epoch achieved this will help us to answer the still open question of the extent to which the early bourgeois mode of production contains not only the possibility but also the necessity of the genesis of classical natural science. The next chapter is devoted to this investigation.

References

Beaumont, O., and Jenkins, J. G. 1957. Farm tools. In Singer Ch. et al., (eds): *A History of Technology*. Oxford 1964, vol. 3, 134–150.

Bernal, J. D. 1971. *Science in history*, 4 vols. Cambridge, MA: MIT Press.

Borkenau, F. 1934. *Der Übergang vom feudalen zum bürgerlichen Weltbild*, Schriften des Instituts für Sozialforschung. Vol. 4. Paris: Alcan. (Repr. Darmstadt: WBG 1971).

Bykov, V. V. 1974. Der konkret-historische Charakter der Verbindung der Wissenschaft mit der Produktion. In *Wissenschaft als Produktivkraft*, 78–118. Berlin: VEB Deutscher Verlag der Wissenschaften.

Clark, G. 1970. *Science and social welfare in the age of Newton*. Oxford: Clarendon Press.

Crombie, A. C. 1952. *Augustine to Galileo: The history of science A. D. 400-1650*. London: Falcon Educational Book.

Daumas, M. R., ed. 1969. *Histoire générale des Techniques*. Paris: PUF.

Dijksterhuis, E. J. 1956. *Die Mechanisierung des Weltbildes*. Berlin: Springer.

Dobb, M. 1963. *Studies in the development of capitalism*, 2nd. rev. Ed. New York: International Publishers.

Feldhaus, F. M. 1930. *Die Technik der Antike und des Mittelalters*. Potsdam: Athenaion (Repr. 1971).

Flanagan, F. J. 1957. Figured fabrics. In Singer Ch. et al., (eds): *A History of Technology*. Oxford 1964, vol. 3, 187–205.

Forbes, R. J. 1957. Food and drink. In Singer Ch. et al., (eds): *A History of Technology*. Oxford 1964, vol.3, 1–26.

Gibbs, F. W. 1957. Invention in chemical industries. In Singer Ch. et al., (eds): A History of Technology. Oxford 1964, vol. 3, 676–708.

Grossmann, H. 2009. The social foundations of the mechanistic philosophy and manufacture (1935). In *The social and economic roots of the scientific revolution (BSPS 278)*, ed. G. Freudenthal and P. McLaughlin, 103–156. Berlin: Springer.

Hall, A. R. 1957. Military technology. In Singer, Ch. et al. (eds): *A History of Technology*. Oxford 1964, vol.3, 347–376.

Harig, G. 1958. Über die Entstehung der klassischen Naturwissenschaften in Europa. *Deutsche Zeitschrift für Philosophie* 6 (3): 419–450.

Hessen, B. 2009. The social and economic roots of Newton's *Principia* (1931). In *The social and economic roots of the scientific revolution (BSPS 278)*, ed. G. Freudenthal and P. McLaughlin, 41–102. Berlin: Springer.

Hobsbawm, E. J. 1999. *Industry and empire: From 1750 to the present day*. New York: The New Press.

Jonas, W. 1969. *Die Produktivkräfte in der Geschichte*. Berlin: Dietz.

Klemm, F. 1954. *Technik. Eine Geschichte ihrer Probleme*. Freiburg: Alber.

Kuczynski, J., ed. 1972. *Wissenschaft und Gesellschaft: Studien und Essays über sechs Jahrtausende*. Berlin: Akademie-Verlag.

Kulischer, J. 1928–1929. *Allgemeine Wirtschaftsgeschichte des Mittelalters und der Neuzeit: Band I (1928): Das Mittelalter. Band II (1929): Die Neuzeit. Mit Register zu Band I–II*. Berlin: De Gruyter.

Kusin, A. A. 1970. *Karl Marx und Probleme der Technik*. Leipzig: VEB Fachverlag.

Lloyd, H. A. 1957. Mechanical timekeepers. In Singer, Ch. et al. (eds): *A History of Technology*. Oxford 1964, vol. 3, 648–675.

Marx, K. 1986. Outlines of the critique of political economy (*Grundrisse*). In *MECW*, vol. 29. London: Lawrence & Wishart (Electric Book 2010).

Marx, K. 1996–1998. Capital. In *MECW*, vol. 3, 35–37. London: Lawrence & Wishart (Electric Book 2010).

Mineur, H. 1970. Mechanik und Astronomie. In *Die Wissenschaft im Lichte des Marxismus (1934)*, ed. H. Wallon. Berlin: Rotdruck.

Mottek, H. 1971–1974. *Wirtschaftsgeschichte Deutschlands*. 3 vols. Berlin: VEB Deutscher Verlag der Wissenschaft.

Naish, G. P. B. 1957. Ships and shipbuilding. In Singer, Ch. et al. (eds): *A History of Technology*. Oxford 1964, vol.3, 582–619.

Olschki, L. 1919–1927. *Geschichte der neusprachlichen wissenschaftlichen Literatur*. Halle: Max Niemeyer Verlag.

Ornstein, M. 1928. *The role of scientific societies in the 17th century*. Chicago: Chicago: University of Chicago Press.

Parsons, W. B. 1939. *Engineers and engineering in the renaissance*. Baltimore: John Hopkins University Press.

Price, D. J. 1957. Precision instruments. In Singer, Ch. et al. (eds): *A History of Technology*. Oxford 1964, vol.3, 582–619.

Ruben, P. 1969. Geschichtliche Bildung und theoretisches Konzept der klassischen bürgerlichen Naturanschauung. In *Naturphilosophie - Von der Spekulation zur Wissenschaft*, ed. H. Hörz et al. Berlin: Akademie-Verlag.

Salaman, R. A. 1957. Tradesmen's tools 1500-1850. In Singer, Ch. et al. (eds): *A History of Technology*. Oxford 1964, vol. 3, 110–122.

Schuchardin, S. W. 1963. *Grundlagen der Geschichte der Technik*. Trans. by G. Ruediger and P. Heute. Leipzig: VEB Fachbuchverlag.

Singer, C. J., et al., eds. 1957. *A history of technology*. Vol. 3. Oxford: Clarendon Press.

Smith, C. S. et al. 1957. Metallurgy and assaying. In Singer, Ch. et al. (eds): *A History of Technology*. Oxford 1964, vol. 3, 27–71.

Sohn-Rethel, A. 1978. *Intellectual and manual labour: A critique of epistemology*. Atlantic Highlands: Humanities Press.

Taylor, E. G. R. 1957. Cartography survey and navigation. In Singer, Ch. et al. (eds): *A History of Technology*. Oxford 1964, vol. 3, 501–529.

Zilsel, E. 1926. *Die Entstehung des Geniebegriffs*. Tübingen: J. B. C. Mohr.

Zilsel, E. 1942. The sociological roots of science. *The American Journal of Sociology* 47:544–562.

Chapter 2
The Development of the Bourgeois Social Formation and the Genesis of Modern Natural Science

2.1 First Section

2.1.1 Duhem's Revolution

Initiated above all by the pioneering investigations of P. Duhem at the beginning of our century, research in the history of science has thoroughly destroyed the idea that modern science rose like a phoenix at the end of the sixteenth century with figures such as Galileo, Descartes, Bacon, etc. as an unmediated contrast to the dark Middle Ages. Research in the wake of Duhem makes it necessary to date the beginning of modern science to 250 or even 300 years earlier and to see the scientific significance of the late Middle Ages in a new light. Furthermore, the importance of this research is not limited to the fact that it reveals the manifold ramifications, interconnections and intercultural dependencies of the genesis of modern science and thus paints this genesis in a new light. At first glance, the most incisive result of this research is that it seems to have conquered the historical dimension of the history of science by questioning the black-and-white opposition between the Middle Ages and the modern era.

We must ask, of course, whether this discourse compensates in the field of the natural sciences for what Romantic historicism accomplished at the beginning of the nineteenth century with regard to law, the constitution of the state, language, etc. In other words, whether undoubted scientific progress is interwoven with an interest in criticizing the unhistorical course of the bourgeois Enlightenment in order to discredit the latter, and, in particular, its sharp distinction between science and non-science. Thus it is undoubtedly no trivial matter for the historiography of science that P. Duhem is to be ascribed to the circle around Avenarius and Mach, who, in view of the crisis in physics at the turn of the century, saw the opportunity to cast doubt on the objectivity of the natural sciences, i.e. to attack the last bastion of bourgeois materialism. At the same time, the bourgeois faith in progress, including with regard to natural science and technology, which had for so long been a source of its

W. Lefèvre, *Theory of Nature and Mode of Production*, Political Epistemology, https://doi.org/10.1007/978-3-032-02295-0_2

own perceived strength, suffered a shock as a result of the First World War. As long as the social form of the development and application of natural science and technology was not subjected to any scientific critique, this shock was bound to undermine the view of the development of modern civilization and its scientific character as a phenomenology of rationality. Instead, modern development was rather viewed as based in the last instance on irrational orientations (e.g. Max Weber), on a loss of sense orientation (e.g. E. Husserl), on repression (S. Freud) or even on oblivion of being (M. Heidegger). Under these conditions, even the Catholic Church could hope either to justify its persecution of modern science as a measure that proved far-sighted[1] in retrospect or—and this is more important for the history of science—to trace the origins of modern science not to anti-Scholastic opposition but to scholasticism, i.e., to set a fox to keep the geese.[2]

2.1.2 *Historical or Philological?*

It was easy for such tendencies to come to the fore in in the historiography of science of our century, because this research usually does not actually follow a historical method, but rather a method that can be described—*cum grano salis*—as philological, partly due to the fact that the philological recovery and development of the corresponding materials and facts was the prerequisite for reconstructing the prehistory of modern natural science. The merits of these philological investigators, among whom can be found scholars such as A. Maier, are in no way to be disregarded; but neither are the disadvantages of the philological method.

The unhistorical idea of the Enlightenment and of nineteenth-century positivism that scientific theories arise, as it were, parthenogenetically, cannot really be overcome by the philological method. With this method, scientific theories—to run with the metaphor—do indeed have fathers, but philologically, spirit can only be recognized as generated by spirit. This, of course, in no way devalues these works, but only points to the narrow limits within which they contribute to reconstructing the genesis of the modern sciences. Far more important, however, are two other concomitants of the philological method. On the one hand, philologically tracing the filiations of items of theory, formulas, terms, terminology, etc., necessarily assumes—whether the researcher is aware of it or usually not—philosophical interpretations of modern science, particularly those of philosophy of science, although these interpretations are often arbitrary and reproduce common prejudices, precisely because they usually remain implicit. On the other hand, however, the

[1] As a popular example of this, see Le Fort (1962).

[2] This seems to me to be the case, at least as a secondary purpose, with some historians of science, such as E. Borchert, B. Jansen, but also with such eminent scholars as P. Duhem and C. Michalski. A—presumably involuntary—contribution to this is also made by Crombie, who sees the suppression of the Averroist interpretation of Aristotle in the thirteenth century as an aid to the emancipation of modern natural science from Aristotle (Crombie 1952, 42 f.); similarly Blumenberg (1985, 346 f).

ordinary use of the philological method contains a concept of development in which the continuity of development is only seen as a gradual modification, while erratic changes are regarded as ruptures. If, as a result, the tendency arises to transform, by selective perception or in some other way, erratic changes whose character of continuity is beyond question in the wake of other knowledge, into steady modifications, not only does this obstruct reconstruction, but it is also possible that real breaks are rendered unrecognizable. With regard to the genesis of modern science, however, it is not only important—as with all historical objects—to distinguish the leaps in which it occurs from the breaks it marks in relation to tradition. It seems to me, moreover, that it is precisely the leaps that contain the key to reconstruction, since they force us to go beyond the level of philology in order to work out the identity in what is different, and to uncover the historical content of the relevant intellectual figures.

These general remarks are well illustrated by the history of astronomy, in which one of these leaps in the genesis of modern natural science will be presented below, the study of which I hope will reveal the deeper continuity that determines the development that takes place in contradictory phases.

2.1.3 *Mathematical Hypothesis or Physical Explanation?*

2.1.3.1 Copernicus

The contradiction has often been pointed out that Copernican astronomy, which under the Kantian title Copernican Revolution is virtually regarded as the epitome of the breakthrough of modern science, in many respects belongs rather to medieval than to modern science.[3] It matters less that precursors of the heliocentric system can be found in the Middle Ages; for these precursors, as the case of N. Oresme shows, could possibly be regarded as early pioneers of modern science.[4] On the other hand, it is important in this context that Copernicus adhered to the Platonic dogma according to which the planets must move only in circular orbits, and to the epicycles of the Ptolemaic system.

This last point in particular is by no means a minor astronomical detail. Rather, the epicycles, together with the assumption of eccentricity and the notion of non-uniform motion of the planets in Ptolemaic astronomy, point to perhaps the most significant peculiarity of medieval astronomy. With these assumptions, Ptolemaic astronomy contradicts Aristotelian physics which, especially with regard to the celestial bodies, had been undisputedly "the" physics since the thirteenth century. Moreover, medieval astronomy in general did not even attempt to bring the Ptolemaic system into agreement with the physical views of its time; the contradiction

[3] Cf. e.g. Blumenberg (1965); cf. also Dijksterhuis (1956, 320 ff.).

[4] Oresme, however, was only concerned with the possibility of the Earth's rotation, not with its movement around the Sun.

apparently did not irritate it.[5] Certain of its usefulness for navigation and calendar calculation, medieval astronomy did not claim to be a form of natural scientific knowledge, but contented itself with propounding mathematical hypotheses which, though in open contradiction to other views of nature, were manifestly useful for practical purposes. With regard to the calculations needed for practice, the Ptolemaic system, which was based on geometric modelling, had an advantage over Aristotelian astronomy, which was a physical explanation. This advantage was purchased by renouncing scientific explanation, i.e. with a certain agnosticism.

Copernicus belongs to this agnostic tradition insofar as he was as unconcerned as the medieval astronomers about the contradictions between his system and the scientific views of his time.[6] His conviction that he was not simply putting forward an alternative hypothesis to Ptolemy's, but that he was truly grasping reality by means of his system, was not primarily based on scientific thoughts, but above all on theological considerations. Copernicus was convinced, on the one hand, of the primacy of the sun over all the other celestial bodies, and, on the other hand, that a simpler system corresponded more closely to reality, because God would not take a detour to put into operation what he could create in a simpler way.[7] (It must be left unexamined here to what extent this last thought, belonging to High Scholasticism, is already to be seen as a result of the assimilating defence against anti-feudal currents of thought in the twelfth/thirteenth century and thus perhaps as an expression of the first stirrings of the modern world view). The gap that separates Copernicus from modern natural science becomes clear when one considers the way in which Kepler and Galileo practised astronomy.

2.1.3.2 Kepler

The difference is not that a man like Kepler would not have been inclined to theological speculation; on the contrary, Kepler's theological speculations often appear darker than those of Copernicus. Kepler can virtually be considered the classic case of how advanced scientific questions and ideas can often result from a combination of solid knowledge and meticulous research with irrational-seeming speculation.[8] And this also seems to apply to Kepler's achievement, which is far more significant than the discovery of the three laws named after him, namely the view that the

[5] Some attempts were made in the thirteenth century, cf. Crombie (1952, 62); but they had no influence on the astronomy of the Middle Ages.

[6] Dijksterhuis remarks (1956, 329) that Copernicus did not close the gap between astronomy and physics, but that his system made this gap a problem.

[7] Cf. Butterfield (1949). Of course Copernicus also used physical arguments, cf. Kauffeldt (1954, 55 f.); but these arguments were neither decisive for his considerations, nor did he try to draw consequences from their (partial) disagreement with official physics (Copernicus advocated the impetus theory).

[8] Crombie (1952, 315ff.); with regard to Kepler's proximity and—over time—distance to Pythagorean speculations, see Cassirer (1940, 263 f.).

planetary system is to be understood as a physical one that can be deciphered with the scientific knowledge valid on Earth. Dijksterhuis, at any rate, writes: " … it is by no means to be denied that the similarity which he (Kepler) sees between the two"—the Trinity of the Christian God and the world structure: sun-Father, fixed stars-Son, moving power of the sun-Holy Spirit—"helped to inspire him to one of his most fruitful ideas, the conception of the sun as *causa efficiens* of planetary motion."[9]

Even if Kepler did not think of the Sun's capacity to act on the planets in terms of gravity but in terms of a kind of magnetism, the crucial point is that he worked on the unity of mathematical representation and physical explanation of astronomical phenomena.[10] This marks his epochal distance from the astronomy of the Middle Ages—and also from Copernicus. Kepler himself was obviously aware of this fundamental difference from conventional astronomy. He formulated it almost as a programme, as already the title of one of his works shows: *Astronomia nova aitiologetos, seu physica coelestis.*

Astronomy sees itself as a physical science; it sees it as its task to explain the causes of celestial phenomena; mathematical modelling and physical explanation are no longer unconnected, even contradictory to each other, but are to form a unity.

The Galileo trial proves that it is precisely the physical dimension of astronomy that could no longer be tolerated by the guardians of the feudalistic world view; at the same time, it makes clear what ideological consequences this physical dimension implied.

2.1.3.3 Galileo's Crime

In the first 70 years after the publication of *De revolutionibus* (1543), the Copernican system suffered no persecution, as is well known, although the Tridentinum falls into this period, and the so-called Counter-Reformation, which opened with it, was by no means concerned with Protestant currents alone. The Copernican system was controversial at this time, but controversial as a mathematical hypothesis. Even when men like G. Bruno and Campanella drew ideological conclusions from the heliocentric system that were directed head-on against the feudalistic worldview, there was brutal suppression of such uses of the Copernican system—G. Bruno was burned in 1600—but the Copernican work was not indexed. It was only when natural scientists like Galileo set about raising the heliocentric system beyond the status of a hypothesis and proving its truth using physical methods, that the Copernican system met with persecution.[11] In 1616 *De revolutionibus* was suspended; that is, it

[9] Dijksterhuis (1956, 340).

[10] After all, Kepler used the term "vis" for this, thus anticipating the second half of the seventeenth century!

[11] Cf. e.g. Galileo's letter to Dini of 23.3.1615.

was not formally placed on the Index, but "only" demanded some corrections, and otherwise set free for further discussion as a mathematical hypothesis.[12]

In the trial of 1633, the question of whether Galileo, in his *Dialogue*, had kept to the requirement to discuss the Copernican system only as a mathematical hypothesis then played the decisive role.[13] Melchior Inchofer, one of the councillors of the Inquisition, clearly names the "offence" for which Galileo is being tried: "The author (Galileo) claims to be discussing a mathematical hypothesis, but he presents it as a physical reality, which mathematicians never do."[14]

What brought the Inquisition on the scene against the attempt to secure the heliocentric system was—if for the sake of poignancy the distinction is permissible for a moment—not primarily the substantiation of heliocentrism, but the attempt at a physical, i.e. scientific, explanation of the world itself. Indeed, this presented a tremendous challenge; for with this attempt the opposition to the feudalistic world view left the level of merely opposing it with another speculation. With this attempt, the ideological struggle of the rising bourgeoisie with the aristocracy, which had been dragging on for centuries, reached a qualitatively new stage: the bourgeois view, as a science founded in reality, opposed a feudalistic worldview that had thus been degraded to irrational speculation.

Thus, the Galileo trial throws a significant light on the widespread view that so-called mathematization is the essential characteristic of modern natural science. Against this background, this view not only proves to be extraordinarily superficial; it also makes it virtually impossible to reconstruct the genesis of modern natural science. The natural science of the seventeenth century is distinguished precisely by the fact that it overcame the speculative use of mathematics that developed in antiquity and reached a new high point in the Renaissance.[15]

---

[12] The permission to discuss the Copernican system as a mathematical hypothesis must, in my opinion, be regarded as serious; cf. Wohlwill (1969, 560 ff.). The demand for correction, however, almost completely devalued this permission, since—be it out of ignorance or malice—it presented the Copernican theses on the central position of the Sun and the movements of the Earth in such a falsified way that any correction, which was unavoidable even for the discussion as a mathematical hypothesis, could be interpreted as a violation of the decree of 1616. The enthronement of Urbán VIII (1623) brought a temporary relaxation of this decree.

[13] Freiesleben gets to the heart of the case—perhaps involuntarily, when he sees it as being less about Copernicanism in general than about the requirement of the decree of 1616 (Freiesleben 1969, 128); the reason for this becomes clear in our context.

[14] Cf. Galilei (1907, Vol. XIX, 348 ff.).

[15] This is also relevant with regard to the so-called anti-Aristotelianism of modern science. Aristotle famously distinguished between the explanatory value of mathematical and physical proofs, i.e. proofs based, among other things, on empiricism (cf. *Post. An.* 79 a, *Phys.* 193 b, *Metaphys.* 1077 b-78 b, etc.). Contrary to popular opinion (for many, cf. Crombie 1952, 83f), this distinction is not one of the core elements of the peripatetic tradition that impedes any progress in natural science; on the contrary, modern science had to adhere to it and did indeed adhere to it (cf. on this Asser 1965; Cassirer 1932; Mielke 1966, esp. 151 ff; Reidemeister 1949; Rietzler 1940; Wieland 1962, esp. 18 f. and 25). The sweeping dismissal of Aristotle's natural philosophy is an example of the common prejudices mentioned above, which are often conveyed in an uncontrolled manner by the philological historiography of science. Cf. generally on this Ley (1966 ff, vol. 2/2, sec. V), and Kristeller (1961).

The third chapter will return to the significance of mathematics for the genesis of modern natural science. This much can already be said here, however: that the relationship of the science represented by Galileo/Kepler to the mathematical and astronomical achievements of the Renaissance involves a qualitative leap; we are dealing with mutually contradictory forms of theory, with speculation and explanatory natural science oriented towards empirical knowledge. Nevertheless, there is undoubtedly a connection between these forms of theory, as is already evident from the fact that Galileo and Kepler saw themselves as Copernicans.

2.1.3.4 Continuity in Discontinuity

The inner connection between these contradictory forms of theory is undoubtedly based on the ideological common ground that transcends their differences. In the work of Giordano Bruno there is even a mediating position in which this commonality is directly expressed. Bruno, who is attached to Renaissance philosophy, namely that of Cusanus, agrees with Copernicus in the speculative mode of thought. But he understands the heliocentric system not as a mathematical hypothesis but as a physical reality. He does not notice in the Copernican system the kinship with medieval theory which strikes us today, but reads it as a shattering of the feudalistic view of the world, corresponding to the revolutionary ferment of his time. Whether he thereby does justice to Copernicus' intentions or not may remain undecided, like the old question whether Osiander's introduction to *De revolutionibus*—the supposedly or actually friendly service of a worldly-wise man—falsified Copernican intentions or nevertheless reflected Copernicus' view to a certain extent.[16] Bruno's ideological interpretation of the Copernican system is probably confirmed by the consistency with which Copernicus articulated the heliocentric hypothesis, which was by no means original. This consistency, which fully justifies the outstanding place that Copernicus holds in the history of science, would have been unthinkable in the case of a man who regarded the heliocentric alternative only as a possibility of thought. In Bruno's favour is also a detail which, with regard to the further development of astronomy, belongs to the weakest points of the Copernican system, namely the rehabilitation, already mentioned, of the Platonic dogma of the perfect circular motion of the planets, which Copernicus upheld against the implicit deviations in the Ptolemaic system connected with eccentricity. For in this rehabilitation Copernicus proves to be a representative of that form of Renaissance thinking that imagined an aesthetic-mathematical world unity, whose thrust against the feudalistic world view probably emerges most clearly in the artistic realism of the Renaissance.[17]

---

[16] See e.g. Koyré (1969, vol. 2, 61 f.).

[17] See Kuhn (1957).—Just as it proved to be proved to conceive of Aristotelian philosophy and modern science solely as opposites, it can be seen here that the mere statement of the Platonic character of Renaissance philosophy can only lead to misunderstandings that make the genesis of modern science incomprehensible. It is important to see that the Renaissance Platonism, which is

The ideological connection between the astronomy of Copernicus and that of Galileo and Kepler should thus no less be overlooked than the fundamental difference between the types of theory they represent. In view of this we must not eclectically emphasize only the similarities or only the differences, or stop agnostically at a dichotomy. If it is unacceptable to interpret as self-deception the fact that Galileo and Kepler themselves understood their work as carrying out the Copernican legacy, and if it is equally important neither to play down the difference between them nor to dismiss the common ground as mere externality, then the historiography of science must grasp the form of theory represented by Galileo/Kepler as a consequence of the form of theory of the Renaissance. That is, it must present the discontinuity contained in the difference as a necessary form of continuity between these stages of the development of modern science.

2.1.4 *The Ideological Thrust*

This task can only be solved by examining, on the one hand, the relationship between the common ideological content of these stages and the forms of theory in which they are articulated, and, on the other, under which conditions the speculative form of theory is necessary and under which it can be abolished.

The Renaissance theory represented by Copernicus and the seventeenth-century natural science represented by Galileo and Kepler are both expressions of the developing bourgeoisie's ideological confrontation with the worldview of feudalism. The main content of this confrontation consisted—according to its philosophical core—in asserting or denying that phenomenal reality is a relation grounded in itself, i.e. an immanent and necessary relation.[18] In order to understand this content in its historical necessity, the reality expressed in the respective positions must be investigated. The debate must thus be studied as a mental reproduction of historical-social relations, and these relations themselves must be made conscious—as tersely as possible—for this purpose.

---

at issue with regard to seventeenth century natural science (i.e. for example Nicholas of Cusa, but not Ficino), in continuation of the pantheistic Plato tradition, which formed the core of the feudalistic worldview. See the following footnote. Cf. Ley (1966, Vol. 2/2, para. V); Kalivoda (1966). On Nicholas of Cusa, cf. above all de Gandillac (1941).

[18] I use the formulation "appearing reality" with regard to the fact that the bourgeois position turned against an objectively idealistic (cf. note 27) world view. In simplified terms, this view knows two realities, the earthly, appearing reality and a spiritual reality that does not appear, at least in its essential content. In order to express that the bourgeois position asserts the former as founded in itself, I characterise it as this "appearing" reality; for the feudalist worldview that was fought against also knew a reality that was founded in itself: precisely this spiritual reality.—On the problems of the formulation chosen here for the philosophical core of the content of the worldview debate, cf. Chap. 4, note 2.

2.2 Second Section

2.2.1 *Feudalism: The Society of Autocrats*

With regard to the worldview fought by the bourgeoisie, the social conditions of feudalism are to be considered. Feudalism is a particular form of social relations, which can be described as the dictatorship of large landed property.[19] For our purposes it is sufficient to emphasize three features characterizing these social relations. First, production under these conditions is private production, the private producers disposing in principle of their means of production; in other words, society does not dispose as society of the sum of the means of production expended.[20] Secondly, under these conditions the production of use-value predominates, while commodity production is found only as an exception.[21] As among all modes of production in history, for the individual producer the society to which he belongs constitutes a condition of existence. However, the way in which, among the two peculiarities mentioned, society constitutes the condition of existence of the individual, this constitutes the third moment, which is especially important for our question, and which characterizes feudalistic social relations.

This third moment appears most clearly in contrast to the bourgeois epoch. In the bourgeois mode of production, the members of society produce and reproduce their social whole through the mutual exchange of the results of their labour; that is, they make, on the basis of private production within the social division of labour, themselves dependent on the production of others. In this way, the social whole is at once the precondition and the result of the private production of the individual members of society, and does not exist separately from their productive life. Under feudalistic conditions, on the other hand, the social relation does not grow out of the productive life of the private producers; for the predominance of the production of use-value means the predominance of self-sufficient private production. The members of society have nothing to do with each other as producers. No longer the original, quasi-natural collective of producers, and not yet a society chained together by the division of labour, feudalistic society appears as a non-society of self-sufficient private individuals; one might almost think that one had discovered a historical epoch in which the individual does not need society for his life. Emphasizing that the class relations of feudalism have economic causes—they are based on the difference between large and small landed property—but are not realised economically only illuminates this peculiarity of feudalistic society from another side. Since labour capacity, means of production, and food are not opposed as the property of different classes, since classes therefore do not represent the fundamental conditions of life against each

[19] I use the term feudalism in this work not only for the time of the developed feudal system, but as a term for the Western form of the dictatorship of large landed property, including the early Middle Ages and absolutism.

[20] Cf. with regard to the bonded peasants Pirenne, 197, Vol. 1, 44 (for original see Pirenne 1963).

[21] Cf. Pirenne (1971, 10 ff.); cf. also Kulischer (1958, Vol. 1, 82 ff.).

other, their relations do not arise and reproduce themselves by interrelations in production. Thus—to put it more pointedly—the social relations of feudalism are not based on exchange, but on the lack of exchange between the members of society, on their lack of relations. Under these conditions, the struggle for the most important and—apart from colonization—non-reproducible means of production, for land, constitutes the core of social life.

2.2.2 *The Hidden Bond of Society*

Nevertheless, if this society did not sink into chaos and civil war but, on the contrary, was able to work its way up from a crude first stage of permanent civil war to a state in which the periodic civil wars were interrupted by phases of relative peace, then this balance in the coexistence of self-sufficient private individuals, which was established at least overall, points to a real foundation of the social whole.[22] This foundation consists in the fact that land could be a personal possession—at the stage of development reached by the productive forces it even had to be divided up, because it could only be worked in individual (family) form—but not private property.[23] Private property in land presupposes in all historical epochs the protection of the territory of all private owners belonging to a society against other societies. Whether private property can exist depends, therefore, in any historical epoch, on the respect for property enforced by and from within society. Where production for exchange, based on the division of labour, prevails, this respect for property is constantly established in the course of productive life; each presupposes for his own existence the existence of others as private owners. This is different when production of use-value predominates. Here the social recognition of a thing as the property of a private individual requires the express approval of the other (or of a certain portion of the other) members of society, thus private property requires its opposite, namely, the guarantee by a collective. This collective reveals itself in this process as the allotting and thus actual owner, who accordingly (as is shown by all forms of fief) attaches conditions to the allotment of property. Unrestricted private ownership of land could therefore not exist in the period under consideration. The disposal of the individual over "his" land was restricted in many ways, and these restrictions show that the most important means of production of this epoch was in the last instance common property; a common property, admittedly, over which society as society had lost all control.[24]

---

[22] This term is certainly an anachronism for the period under consideration. For it presupposes a society in which the community as a sphere of civil relations is distinct from that of intercourse between communities, which certainly cannot be said of the Middle Ages. On the other hand, it is a question of struggles in which society is formed, which therefore have an internal character. I am not aware of a term that expresses the contradictory character of these struggles.

[23] Cf. Hegel (1991, § 62); von Schwerin (1944, 184); Mitteis (1958, 625 pass.).

[24] On the restrictions, cf. von Schwerin (1944, 187 ff.); Mitteis (1958, 500 ff., 531 pass. and 591 pass.).

The contradictory character of ownership of the most important means of production—in the contradiction, namely, that it was the property of a society which could not behave towards it as a societal owner—explains why historical research did not discover for a long time that feudalistic society had its inner bond in the common ownership of land, and therefore sought this bond in the ideals and views of the time.[25] Furthermore, since the coherence of this society was based on this contradictory property character of land and soil, we are also confronted with the reality whose necessity and coherence could only be understood in a sphere separate from the appearing world.

2.2.3 *The Social Bond as an Otherworldly Entity*

The social whole was based on the common ownership of land; the consequences of the contradictory form of this common ownership, however, was that the appearance of this social whole displayed, at best, mutually indifferent, self-sufficient private individuals and—in the worse, frequently occurring, case—the struggle of private individuals for land. While the violent attempts from above to transform land into unrestricted private property—together with their consequences for the existence of society—dominated the form of the social whole, the foundation of the social context, the common ownership of land, found expression in it only in the fact that these attempts at privatization ultimately failed.[26] The balance found in the feudal system, however much the common property of land was its real basis, could not be regarded as a form of this earthly foundation of society; to the lower estates it appeared arbitrary, to the higher as a truce to be maintained as long as one is too weak for war. The form of the social bond showed it to be a bond without inner necessity; that in which it was founded, land as common property, appeared not as common property, but as its opposite, as private property with impediments. The form of the social bond, therefore, not only did not quite coincide with its essence, but presented it as its opposite. If, in view of the open or latent civil war permanently endangering the existence of society, it was in the interest of both the rulers and the ruled to envision the social bond, then this vision could in no way ignore this objective contradiction, existing independently of consciousness, between the appearance and the essence of this bond. If it could not locate the essence in the appearing reality, then only one way remained for it to be able to hold on to it at all: it had to conceive of the essence of the appearing reality as independent of and superior to this appearing reality, as an extra- and pre-worldly essence; i.e., this realization could only be accomplished in the form of objective idealism.[27]

---

[25] Cf. e.g. the literature report in Kulischer (1958, vol. 1, 29–32).

[26] Cf. Pirenne (1971, 11).

[27] I use the distinction between subjective and objective idealism as F. Engels developed it with respect to the recognisability of reality. His distinction between idealism and materialism (Engels 1962, 275f., 1990, 366f.) also plays a guiding role for my terminology. This terminology has its

The necessity of mentally representing social reality in this form is perhaps most easily revealed by the attempt to uncover reasons for this form of mental representation in the situation of the peasant class as well; that this visualization corresponded to the interests of the large landowner class is probably obvious.

2.2.4 The Peasant Perspective

The emergence of land as an imperfect form of private property, which consigned the realization of the social bond to heavenly spheres, was based on the contradictory manner in which land was common property at this stage of development; this contradiction was in turn founded in the individual cultivation of the land, which corresponded to the stage of development of the productive forces. The social form of labour not only necessarily produced private possession of land, but also differences in wealth among the autarkic private producers. The oppression of the small by the large landowners and the wars of robbery among the large landowners were the inevitable result. A mental representation which not only reproduces the contradictions of this social reality in the manner of objective idealism, but grasps these contradictions, would have to recognize private labour and private possession as the core of the contradictions. But how should this realization be possible where the state of development of the productive forces did not permit any other social form of labour, and—perhaps even more important—where the peasant in this realization

primary reference to philosophical theory, which is certainly a problem with regard to the worldview of feudalism, which is religion and, as theory, theology. In what follows, the worldview that places God as spiritual reality before and above appearing reality is included under the term objective idealism, even though where I discuss medieval theory, I treat Christian theology as philosophy.—For the period under discussion here, the term objective idealism refers first and foremost to Platonism, both to the dualistic idealism found in the Augustinian branch of Neo-Platonism (cf. on the characteristics of this branch Chenu 1969, 268–316, esp. 280 ff.) and—as I believe—finds its most effective form in the Aristotelian Thomas in the criticism of the twelfth and thirteenth centuries (cf. Hirschberger 1969, esp. 63 ff; Grabmann 1948, 45ff.; Santeler 1939, 146ff. and 201 ff.; H. Ley 1966, vol. 2/2, Sect. IV/1), as well as to pantheistic or monistic Platonism, which, as the other branch of Neoplatonism, had an impact on the West and lived on there (on the one hand via Pophyrios, the Calcidian 'Timaeus' and Boethius, on the other hand in its Greek-Byzantine further development via J, Scotus Eriugena and later via a direct reception of Pseudo-Dionysius Areopagita, thirdly via its Arab further development, primarily via Avicenna. Cf. Chenu 1969; Koch 1969). Secondly, however, the parts of Aristotle taken up by medieval philosophy are also in mind, in which Aristotle—despite his intention—does not fundamentally differ from Plato (cf. among others *De An.* II 5 or the late pieces of the *Metaphysics*; cf. on this Düring 1966, esp. 46ff.). The differences and contrasts within the objective idealist camp alluded to here will be discussed in part in later contexts.—The philosophy of antiquity could, of course, become the philosophical foundation of the feudalist world view, because it was formed under social conditions that corresponded to those of the Middle Ages despite all the differences in the essential features (cf. above p. 54 f.) (cf. Lefevre 1971, esp. note 30 to p. 70 and the literature cited in note 23 to p. 65).

would have had to negate himself as a private owner?[28] It cannot be surprising, therefore, that even in the peasant revolts of the early sixteenth century, which were distinguished not only by their extent but equally by the clarity and decisiveness of their aims, the struggle was directed against the oppression of the small by the large landowner, not against the root of this oppression, against private property.[29] Republican equality between large and small landed property, this is the slogan of the great Peasant War of 1525, which probably expresses in this slogan the quintessence of all the peasant revolts of the Middle Ages in a developed, namely political, form.

The realization that the contradictions of feudalist society are founded in private labour and private possession was unavailable to the peasant class for objective reasons: this is why the root of aristocratic rule remained incomprehensible to the peasantry. Aristocratic rule also derived no semblance of necessity from the role it played in the productive life of society. For the aristocracy, unlike capital, had no function in productive life (which throws light on the aristocracy's efforts to make itself stand out as militarily indispensable). Incomprehensible in its origin, incomprehensible from its function, and yet at that time an apparently unalterable reality, that is, a reality with no discernible earthly foundation, and yet apparently well-grounded in reality—how could the mind to come to terms with this reality of the aristocratic class and thus of the class division of society, without yielding to the prevailing objective idealist worldview?

The need to idealistically reproduce the social conditions of feudalism in the mind does not, however, also automatically imply the possibility of universalizing this mental representation of social reality into a worldview. This requires, after all, that nature also be conceived idealistically. But how is this possible, since the conception of nature cannot be formed in complete contradiction to the conception of nature implied in material production, and it must seem inconceivable at first glance that even a single field could be cultivated on the basis of an objective idealist conception of nature.

[28] In the West, peasants were generally private producers who owned their means of production (including land), despite serfdom and noble titles of ownership. This only changed with the so-called second serfdom or with the eviction of peasants from land and soil, cf. footnote 21 in this chapter.

[29] Thomas Müntzer's propaganda, which contradicts this statement, cannot, in my opinion, be regarded as an expression of the aims that guided the mass of the insurgents, precisely where it anticipates the need to eliminate 'private property'. It should be noted, incidentally, that Müntzer's thought is determined by the more developed form of social contradictions that was evident in the cities. Cf. Bensing and Rüdiger (1973, 27–29, 33); cf. Lenk (1974); cf. also Smirin's differing view (1952, 314 ff.).

2.2.5 *Worldview: Image of Nature—Form of Production*

In this epoch, agriculture and animal husbandry, i.e. a mode of production in which man uses the biological production processes of nature, predominated in material production. In view of this fact, it is inconceivable that natural reality could have appeared to man as something incoherent. It is conceivable, perhaps, that at the stage before the first horticulture and animal husbandry, when people merely appropriated natural things, when the economy of occupation, not of production, prevailed, nature may have appeared to humans predominantly as an incoherent collection of things: the discovery and use of nature's production processes may have been the first and decisive step in being able to conceive of nature as a whole.[30] The stage of production considered here, however, built on a long history of conscious use of natural processes of production.

2.2.5.1 Nature as Subject

We must first agree on what characterizes the use of biological production processes at this stage, in contrast to the use of the productive potencies of nature in later industrial production. Humans find such biological processes of production that in principle produce adequately for the needs determined by their bodily constitution (here the difference, for instance, to electricity, which is used as a production potency in the industrial age). The reason for this character of certain natural biological production processes is obvious: humans could develop biologically only in the milieu of this biological production of nature. Fauna and flora were our natural-historical conditions of development and preservation, and in this respect they themselves created humans as natural beings to whom their production is tailored. Accordingly, humans do not have to take these natural potencies out of their natural sphere of action—as, for instance, in the case of electricity—and/or let them take effect in other contexts, in order to be able to satisfy their needs by their results; rather it is sufficient to promote and modify these natural processes in their existing sphere of action. They correspond to his purposes without being set in motion by purposeful human activity. These processes do not, of course, take place in order to produce anything, and certainly not in order to produce anything that people can use.

We take this for granted, but we must ask whether the *quid pro quo* of a finalistic conception of this natural production was not self-evident under the conditions of the stage of production at that time. That the interpretation of the coherence of nature as a coherence of production must be thought of as a projection of human production, together with its own teleological structure, onto nature, may be regarded as unprovable speculation, however much the animistic magic of the

[30] These terms were coined by the archaeologist Jaroslav Böhm; cf. Sellnow (1961, 107 ff.).

hunting peoples, for instance, speaks in its favour.[31] For our purposes, however, the matter can be satisfactorily explained by the fact that at this stage of development the utilization of biological processes of production in nature does not reveal this natural production as a moment of human labour, but, conversely, human labour appears as a moment of natural processes of production. The natural production process utilized does not follow from the purposively determined human activity, which merely promotes a process of nature operating of its own accord; the result of this process corresponds, rather, to the purposes of human activity and is not set by them.[32] It is therefore impossible to see how a distinction can be made in this activity between *causa efficiens* and *causa finalis*. In my opinion, this distinction can only be made in an activity that abolishes the existing natural connections between the natural things it uses and establishes between them a new connection determined by the purpose of the activity.

In artisanal production, without which no agriculture is possible—irrespective of whether it is already separated from agriculture on account of the so-called second great social division of labour, or whether it is still mainly performed by the agrarian producer himself—there is such work, which permits a distinction to be made between *causa efficiens* and *causa finalis*. This can be seen in the simple use of some natural thing as a tool. By negating the natural context in which this natural thing arose and was preserved (for example, by breaking a branch from a tree) and inserting it into another context determined by human purposes (for example, by using the branch as a lever), human first show that they do not consider the natural context of the natural thing to be necessary. In the context determined by human purposes, in which humans seek to use it as a means, we experience the determinant contribution of the natural thing; but at the same time we experience—not least on account of the "spite of the object"—that the context determined by human purposes is external to the natural thing. We cannot possibly regard the result of labour as something that the natural things used as means of labour would also have wanted of themselves. The contribution of the means of labour is clearly separate from the contribution of purposeful human activity, and this separation makes the distinction between *causa efficiens* and *causa finalis* not only possible, but necessary.

2.2.5.2 Nature as Quarry

Artisanal production confronts the purpose-determined and thus necessary context created in labour with a nature whose context is made perceptible by this labour as precisely non-necessary; if it does not appear in the light of this labour even as a

[31] In my opinion, more reliable statements in this field could be made possible by research that reconstructs the prescriptive phase of human thought in cooperation with the necessary ontogenetic (genetic epistemology, etc.) and phylogenetic (archaeology, ethnology, etc.) investigations. As far as I can see, this has not yet been attempted anywhere in a promising form. Translator's note: cf. the text "Between two natures" in this volume.

[32] On the use of the term "set" in this context, cf. Lukács (1980).

mere conglomerate of materials and raw forms for means of labour and objects of consumption, as it were as a sole quarry. The necessary interrelationships of nature, from which this inexhaustible arsenal of useful things results, cannot be used for artisanal production and thus cannot be experienced.[33]

Artisanal production naturally makes use of the necessary causal relations of nature by employing natural things as means of labour; the use of these natural things consists, after all, in exerting through them an effect on the object of labour that corresponds to the purpose of labour, an effect that originates in the nature of the means and object of labour. In this respect a natural process of production takes place between materials and means of labour on the one hand, and the objects of labour on the other, and in this respect artisanal production also makes use of natural processes of production. But this use is not only specifically characterized by the fact that these natural work processes must first be set in motion by human activity, because they do not take place of their own accord; it is also characteristic of this use that it understands how to make use of the formative potentials of nature that are active in these processes only as properties of natural things.

Thus the use of natural processes in artisanal production differs from that in agrarian production in that the natural processes of production it uses do not take place of themselves and without the intervention of humans as tailored to the needs of humans; but it is equal to agrarian production in that it can make productive potencies of nature work for it only as properties of natural things. In this, artisanal and agrarian production differ at this stage from later industrial production. Like artisanal production—and unlike agrarian production—industrial production cannot use the formative potencies of nature in production processes that naturally take place in a manner appropriate to human needs. But it differs from artisanal *and* agrarian production of the pre-industrial stage in that it does not use formative potencies of nature as properties of natural things. Rather, on the basis of insight into the general form of these potencies, industrial production constellates natural things in such a way that these potencies come into effect in a process appropriate to human purposes (cf. "Forms of application of natural forces" above).

Now, the fact that natural potencies are utilized in artisanal production, on the one hand, only as initiated by human activity and, on the other hand, only as properties of natural things, makes it impossible to regard the contribution of the means of labour as utilized natural production at all.

In fact, in this work, it is not the natural things employed that enter into necessary functional relations to one another according to the laws of nature, but merely

[33] As I am formulating this very pointedly, a possible misunderstanding must be prevented. If, from the point of view of craft work, the grain appears as a "conglomerate of materials and raw forms", this does not mean that it appears as unstructured material, as a mere substrate of possible shaping (like certain varieties of subjective idealism). Rather, it is always and necessarily an arsenal of certain things distinguished by certain properties and forms. Moreover, the craftsman naturally cannot do without knowledge of natural production processes; but this remains marginal for him and cannot justify a conception of the whole of nature as a connection of inner necessity founded in itself.

moments in them, which could be worked out by practice at this stage only to a certain degree, but not be made available independently of the respective natural thing. Although the contribution of the means of labour employed is a necessary action of natural law, it cannot be regarded as something in which the necessary connections of nature reveal themselves. It is after all about a necessary connection only between moments of the natural things used as means of labour, while, in the work process, there is no connection necessary to nature between these applied natural things themselves, but only one set by human will. Only when practice is able not only to set these interrelationships into activity as properties of natural things, but also to bring them forth from their general and necessary conditions in the desired form, can they be recognized at all as natural formative potencies and, in them, nature in its necessary interrelations.

2.2.5.3 Indifference of Worldview

Artisanal production, then, presents the natural whole to which it owes its means of labour as a connection without necessity, by taking from it the necessary natural things; but the connection which it establishes through labour between these natural things cannot appear to artisanal production as another form of the natural connection of these natural things, but only as an arbitrary connection enforced by the purposes of labour.[34] The attempt to overcome the finalistic conception of the natural whole, which corresponds to the use of natural processes of production in agrarian production, cannot find a foothold in artisanal production, since in it, nature does not appear as a necessary whole at all.[35]

Indeed, we must consider whether it is sufficient to say that the experience of nature in artisanal production cannot form a counterweight if the mental representation of social reality under the dictatorship of large-scale landed property—in its universalization into an objective idealist conception of the world—relied on the finalist conception of the whole of nature prefigured by the experience of nature in agrarian production. Perhaps it can be argued, moreover, that the dualistic formation of the objective idealistic world-view, which is to be regarded as an apologetic type of the objective idealistic form of mentally representing the social reality of this epoch, has a foothold precisely in the experience of nature in artisanal production. The finalistic conception of the whole of nature, which corresponds to agricultural production, could in itself only serve a pantheistic form of objective idealism, which

[34] At the same time, it may become clear how foolish it is to look down on the Aristotelian distinction between natural and forced movement as a lack of abstraction in everyday ideas, which was an obstacle to the development of modern physics. This distinction stems from a certain stage of development, not of the intellectual but of the practical potency of man vis-à-vis nature, and can therefore not be overcome without expanding the practical power of man. Not to give it a thought suits people who, from their date of birth alone, are certain that they can treat Aristotle as a "dead dog."

[35] Cf. note 33 in this chapter.

can even come very close to speculative materialism and which, in any case, is to be understood as an oppositional type of the objective idealist form of mentally representing the social reality of this epoch. The experience of nature in manual production, on the other hand, in which nature does not appear as a necessary context, can rather underpin a worldview according to which the inner necessity of the whole cannot be found in the appearing reality, but in a transcendent sphere of being which relates to it in the manner of a primordial image, in front of which the appearing reality can only be regarded as a contingent context.

Whether this assertion stands or not, it seems impossible to regard the conception of nature implied by artisanal production as a potential against an objective idealist view of the world. It turned out, however, that this view contains the distinction between *causa efficiens* and *causa finalis*. A conception that conceives of the appearing world as a connection of immanent necessity can tie in with this. However, this conception cannot come from the experience of artisanal production, but must come from the mental representation of changed social conditions; the image of nature in artisanal production perhaps contradicts this conception as little as the objective idealist view of the world; but this conception can in no way appeal to an experience contained in artisanal production in which nature appeared as a necessary connection. It is because of the undeveloped nature of material production that the conception of nature implied by it provides no point of reference for a thought directed against the objective idealist conception of the world. The incompatibility between artisanal production and the application of scientific knowledge in material production, which was the subject of the previous chapter, presents itself in the context of this chapter in the fact that the conception of nature implied by artisanal production is not able to support a scientific worldview. This confirms that as long as scientific knowledge cannot be applied to material production and thus cannot be proven, the conception of nature is decisively determined by the mental representation of social reality. On the other hand, this mental representation is determined by the practical attitude to nature at this stage, inasmuch as it must reflect the lack of development of the conception of nature contained in this practice: humans are only as powerful vis-à-vis reality in thought as we are in practice.

2.2.6 *Reified Thinking*

The core of the lack of development of material production at this stage consists in the fact that it can only apply the formative potencies of nature as properties of certain natural things. This also erects the essential barrier to the mental reproduction of reality at this stage. As long as the lawful interrelationships of reality can only be grasped and thus also be comprehended in their appearance as properties of things, thinking can only attempt to grasp the essence of reality in the essence of things, and thus reify the essence. Here, in my opinion, lies the reason for the preponderance of a merely predicating thought in this epoch, the metaphysical character of which

Hegel criticized in the "First Attitude of Thought to Objectivity."[36] This thinking necessarily entangles itself in irresolvable difficulties because it attempts to comprehend the essence of things detached from the relational and developmental context in which they stand.[37] The best minds, which saw these difficulties and wanted to escape them, could, of course, only postulate and construct relational connections speculatively. It is precisely in such outstanding speculative achievements, however, that the limit drawn to thought by the undeveloped nature of material production becomes clear. To illustrate this, it may suffice here merely to refer to a characteristic of the concept of essence in Plato and Aristotle.

Both saw the inevitable weaknesses of a metaphysical concept of essence, in which essential properties, separated from the individual and accidental properties of the individual specimens by an abstraction concentrating on commonalities, are predicated of a substance. Apart from the problems posed by the separation of properties on the basis of such abstraction, this metaphysical concept of essence cannot satisfy, above all, because this essence lacks all necessity. It is, after all, only a compilation of properties that badly and fairly reproduces the factual constitution of real things, to which, by being predicated of a substance, is merely added the assertion that they stand in a necessary, essential relation to one another. It is about overcoming this deficiency in the metaphysical concept of essence, that Plato and Aristotle are concerned, when, with regard to the ideas and/or ousia, they operate with pairs of opposites which, in their complementary character, permit the essence to be comprehended as an essential relation, as a necessary connection.[38] But this solution, proving the rank of these philosophies, which is found for the essence of essence, does not, of course, remove any of the difficulties mentioned when it comes

[36] Cf. Schenk (1973, 115 ff. and 353 ff.); cf. Hegel (1970/2010), §§ 26–36; on the term 'metaphysics'. cf. esp. §§ 27 and 32.

[37] In my opinion, this is not contradicted by the fact that things are conceived as being in relation to the universe and each of its members. For as long as this relational structure is not conceived in terms of the unity of the properties of things and the lawful form of the actual relations of action between things, but, conversely, as long as these relations are conceived in terms of a perfect being conceived as a thing in itself, the essence of reality remains a thing-like essence.—If, from the fourteenth century onwards, the tendency can be observed to place the question of the reality character of the relationship contained in predication at the centre of the investigation, then it would have to be examined whether or not this can be seen as an expression of bourgeois development.

[38] As far as I know, Plato discusses this problem most extensively in the *Sophist* (245–259, esp.: 248a4-249d5, 251a8-252dll, 253c6-254b6, but above all 254b7-255e7); there he works with the pairs of opposites στάσις-κίνησις and τούτο-τοῦτερον. In Aristotle, the corresponding sense is found in the doctrine of the ἀρχαὶ, which is related to the first category (cf. *Metaph.* 1070bl3-1 5); the attempts preceding the pair δύναμις-ἐνέργεια to contrast the εἶδος with the στέρεσις show this (cf. especially *Metaph.* Book A) perhaps show most clearly that he is concerned with a self-referential connection (cf. in parallel to this the explanations given by Wieland 1962, 88 ff., esp. 89, on the pair of terms καθόλου-καθέκαστον).

to the formation of specific concepts of essence.[39] It remains powerless in the face of the problems of such concepts of essence, because these problems do not stem from a lack of capacity of thought, but from practice.

2.2.7 *Speculative Elevation*

The autonomous overcoming of these problems by thought can only consist in rising in the mind above the limits of real, practical conduct to reality. If, in a situation in which the stage of development of material production does not allow lawful interrelationships to be controlled and in which, therefore, it is not possible to prove knowledge of these interrelationships, but thinking does not want to refrain altogether from forming an idea of these interrelationships, then it is relegated to the level of speculation. The positive character of this level of development of material production, which we have illustrated above in Kepler, does not, however, attest to the impotence of thought alone. Rather, the price of this elevation is to renounce any role in overcoming the limits set for the thinker by the state of practice.

As a rule, however, this renunciation is not even perceptible as a renunciation under social conditions which make theory the privilege of a class living in leisure. For antiquity there is not a word to be said about this. The remoteness from practice which is characteristic of the theory of antiquity—apart from exceptions such as Archimedes, Heron of Alexandria, Pliny Jr. and Poseidonius—undoubtedly does not apply to the Middle Ages to this extreme degree.[40] An important reason for this is probably to be seen in the fact that medieval production was not based on slave labour and therefore material labour was not considered a disgrace for a free man as it was in antiquity.[41] Nevertheless, even in the Middle Ages theory and material production had little to do with each other.[42] The state of development of material production is—as we have seen—one essential reason for this. Another, however, is that in the Middle Ages, too, the preoccupation with theory as a rule depended on membership of one of the two estates, which—apart from a few religious communities—did not work.

Every theory directed towards the coherence of reality had to have a speculative form at the stage of development of material production at that time. The position of theory, however, far removed from practice as a result of the social conditions under the dictatorship of large-scale landed property, indicates that the speculative form has a specific affinity for the mental reproduction of precisely these social conditions. The content of this mental representation consisted in grasping the essential

---

[39] It cannot be discussed here whether or not such philosophical achievements can be explained without the brief flourishing of the bourgeois mode of production in the Greek city states.

[40] It is certainly significant that Heran's technical inventions usually had nothing to do with material production, but were used for the entertainment of evening parties. Cf. Kuczynski (1972, 52–54, 70).

[41] Instructive documents on this attitude can be found in Klemm (1954, 5–11).

[42] Cf. Crombie (1952, 143ff.).

connection of the appearing reality in a sphere transcendent to the appearing reality and preordinate to it, and in this way to represent the appearing reality at the same time as void as well as unchangeable by earthly measures; speculative thought is without doubt the adequate thought form for this content. It is to be assumed, however, that speculative thinking can only serve a content that is the mental representation of another social reality as a form that contradicts it.

2.3 Third Section

2.3.1 Commodity Production: The Appearing Bond of Society

The objective idealistic conception of the world represents the necessary mental processing of the social conditions under the dictatorship of big land-ownership, and its overcoming therefore depends on the change of this social reality. The development of the bourgeois mode of production changes this reality. In our context, it is important to examine how this development changes the social reality of feudalism precisely in those traits that were decisive for the formation of the objective idealist world view. For the mere listing of such data as the number of inhabitants of cities, trade routes, forms of business, etc., is in itself unable to contribute anything to the answer to our question.

The bourgeois mode of production, like the feudal mode, is production of private producers. It differs fundamentally from the feudalistic in that the private producers, on account of the developing division of labour, regularly or even already predominantly produce for exchange. This difference also changes the way in which society is a prerequisite for the existence of the individual. The advance of commodity production at the same time dissolves the real foundation of social coherence under the dictatorship of large-landed property, the common property character of land, and gives rise to a new social coherence: the dependence of each private producer on the other private producers, based on the division of labour. If the old social whole was a condition of the existence of self-sufficient private individuals between whom there was no connection in productive life, the new social bond is tied precisely by the connection which exists between private producers on account of the division of labour and the exchange of products in productive life. Production for exchange, based on the division of labour, is a form of social production, however undeveloped. The social reality founded on this form of social labour does not require its mental representation to seek its bond beyond its appearing reality; it presents its bond in the exchange of commodities. The coherence and unity of society no longer constitute a problem that can be "solved" only theoretically; they have become a question answerable through practice. The creation of the nation and the national market, its deepening within, its expansion outward to the world market – these are answers to the question of the coherence and unity of social reality.

This coherence appearing in social reality undoubtedly made possible the development of a thinking directed against the objective idealistic conception of the world, and indeed it arose everywhere with the development of commodity-producing society. Its development, however, is extraordinarily contradictory and poses difficult problems for reconstruction, of which I can only take up the most important for this work, namely the question of the extent to which the contradictions of the bourgeois mode of production are already reflected in the contradictory development of this thinking, or, in other words, the extent to which thinking based on the bourgeois mode of production, given the contradictions of this mode of production, can be of any significance at all for the development of scientific knowledge.

2.3.2 *Contradictory from the Start*

The social relations arising with the bourgeois mode of production are contradictory from the beginning. The bourgeois mode of production is based on private production; consequently it knows no social disposition over the means of production used in society, thus no social disposition over the social production that has come into being with it, and thus also no control over the social relations resulting from this social production. It is true that, unlike the mode of production under the dictatorship of large-landed property, it is social production, but one which the members of society do not control. The social whole that has appeared is an unmastered context.

2.3.3 *Stages of the Contradiction*

What the contradictory character of the bourgeois mode of production means with regard to the development of scientific knowledge cannot be judged without considering the historical stages of development of this contradictory character of the bourgeois mode of production. Nothing could be more confusing and detrimental to reconstruction than to impute the unfolded capitalist form of this contradiction to the early phase of bourgeois society. It is true that in exchange-value it is abstracted from the outset whether the producer also has at his disposal the material conditions of labour; exchange-value thus already "imputes" or "presupposes" (*unterstellt*), as Marx puts it, the developed form of the contradiction of bourgeois society, namely, the separation of labour-power on the one hand and means of production and subsistence on the other.[43] But the bourgeois mode of production does not begin with this most developed form. Rather, its contradiction appears at the beginning as the opposite of this most developed form: the private individuals producing for exchange dispose, as a rule, of their material conditions of production.[44] At the beginning of

[43] Marx, *Grundrisse*, chapter on Capital, section 'Forms Preceding Capitalist Production', in Marx (2010, 433).

[44] Kuczynski insists that the evidence of wage labourers in the early stages of bourgeois development (up to and including the fourteenth century) should not be interpreted in such a way that the relationship between wage labour and capital should be regarded as a characteristic form of the bourgeois mode of production (Kuczynski 1961, vol. 22, 3).

the bourgeois mode of production, it is not the separation between labour-power and the material conditions of production, but the dispersion of the means of production as the property of individual producers that makes social relations unmanageable. This pre-capitalist stage of the bourgeois mode of production lasted—apart from Florence and Venice—at least until the sixteenth century, although it should also be noted that until the eighteenth century the capitalist mode of production did not dominate even in countries such as Holland and England.

2.3.4 *Early Bourgeois Forms of Opposition*

More precisely, then, it is necessary to ask what the contradictory character of the bourgeois mode of production, in the undeveloped form that it had in its pre-capitalist stage, meant with regard to the development of scientific knowledge. This, again, as it seems to me, cannot be judged without considering the fact that the contradiction of the bourgeois mode of production, as yet scarcely developed in the pre-capitalist stage, was covered over by the contradiction which this mode of production formed with the feudalist one. It must be remembered that the bourgeois mode of production arose in and under the dictatorship of large-landed property; it thus formed a contradiction to the feudalist mode of production and, on the other hand, required it for a long time. The contradiction between these two modes of production thus in its turn shows phases of development which can be distinguished. In my opinion, this is the key to the solution of not a few problems facing the reconstruction of the prehistory of modern science, as can be illustrated with just a few sketchy strokes.

2.3.4.1 The Struggle for Urban Autonomy

Thus, in my opinion, it loses all astonishment that the struggle of the urban bourgeoisie for urban autonomy was articulated almost exclusively in objective idealistic form, if one takes into account that, on the one hand, the social content of this struggle does not differ in anything from that of the struggle of the peasants against levies, *Fron*, etc., and that, on the other hand, until the twelfth century, cities and their immediate surroundings only exceptionally represented a micro-society based on commodity production.[45]

[45] The heresies of the twelfth and thirteenth centuries show what opposition potential could be gained on the basis of objective idealistic views. The Cathars (as well as the Bogomils in the Byzantine Empire) show that even the dualistic form of objective idealism can be turned into opposition, whereby the point of all Manichaeism is that instead of the opposition perfect God: sinful world, it sets the opposition good God and (in the followers of the heresy—ideally or practically—realisable) good world: evil God and evil world. Of course, not all Manichaeism leads to an action directed towards this world as in the case of the Cathars, in whose movement, of course, the most

2.3.4.2 Nominalism and Latin Averroism

In general, only long-distance trade had a developed form, while the domestic structure still showed production for barter only as a supplement to the dominant production of use-value. From about the twelfth century onwards, in some areas of Europe (southern England, north-western and southern France, southern, western and northern Germany, and above all in northern Italy) a first flourishing of the market developed, whereby it was decisive that agricultural production in these regions also participated to a certain extent in the production of commodities. It is in these regions that the new social whole first appears and finds its intellectual expression in the first critique of objective idealist thought (nominalism and Latin Averroism).[46]

---

diverse social elements worked together. For our context, it is interesting that after the first great crushing of this movement, the urban bourgeoisie of southern France continued and led it, giving it a direction that allowed it to find connection with the nominalism of heterodox Averroism and its 'double truth' (cf. Ley 1966, vol. 2/2, para. 11/4). In the Begards and Beguines, on the other hand, we find the pantheistic form of objective idealism as the basis of their opposition, i.e. the front against the dualism of the feudalistic worldview.

[46] For early nominalism, evidence that it is to be interpreted as an expression of bourgeois-antifeudalist opposition is at best very indirect. With regard to Roscellin, perhaps the only evidence is that Anselm of Canterbury is said to have mobilised parts of the population against him, which speaks against a controversy conducted only among intellectuals. With regard to Abelard, there are more significant indications: his unusual attraction as a teacher; the veneration of Heloise by the people; the lenient condemnation by the Council of Sens, out of all proportion to the accusation, which suggests that strong anti-clerical social forces worked in Abelard's favour, and so on. (cf. especially the lecture given by J. Le Goff at the 12th meeting of medievalists in Cologne in 1962; cf. also Le Goff's work cited below). Sidorova (1959) points out that a distinction must be made between Abelard's subjective intentions and his objective effect; with regard to the objective effect, however, there can be no doubt that Abelard's teaching was taken up by the bourgeois opposition movements of the twelfth century. This is proved above all by Arnold of Brescia; this follower of Abaelard was not only the leading theoretician of the Roman republic that arose from the revolt of 1143 (cf. Kosminski et al. 1958, vol. 1, 329 and 497 f.), but also his political turn towards the abolitionist movement. But his political twist on Abaelardian nominalism remained effective right up to Cola di Rienzo and Jan Hus. For the School of Chartres, it should also be noted that its outstanding representatives (e.g. William of Conches and Gilbert de la Porée) were accused of heresy, but not treated accordingly. The circumstances of the trial of Gilbert at the councils of Paris and Reims (cf. Gammersbach 1959) indicate that the question of a strong feudal central power, which was usually fought over between kingship and church, had its inner-church counterpart.—In this context, it is generally important not to be irritated by the fact that the bourgeois opposition at this time was not directed against feudalism as such, but primarily against clerical feudalism, its economic practices, etc. The Church was the most important exponent of feudalism. The Church was the most important exponent of the separatist and disintegrative tendencies of the feudal nobility, which repeatedly disintegrated the still barely existing social unity, and against which the bourgeoisie supported the royalty—without always meeting with opposition, as e.g. the behaviour of Emperor Frederick I towards the Roman Republic of 1143 shows. Amalrich of Bena is perhaps the best example of this. How the Church was most affected by these centralising efforts the Church was the most affected by these centralising efforts, so it was its strongest opponent. It is in keeping with this that the development of criticism of the secular church and the demand that the church be limited to its pastoral tasks can be traced from Roscellin to Amhold of Brescia and his followers, a tendency that had its roots in parts of Franciscanism of the 13th and 14th centuries (above

This cannot, of course, mean that from now on all articulation of opposition in objective idealist form must cease at a stroke. The bourgeois opposition remained embedded in a camp of anti-feudalist opposition composed of the most diverse social elements (peasants, urban plebeians, parts of the nobility who—as, for example, in southern France Raimund VI of Toulouse—attempted to transform separatism into a nation-state direction). The unifying bond of this camp is the fight against feudalism concentrated in the rejection of the secular church, while the heterogeneous parts of this camp develop this (often much broader) struggle from different, partly mutually exclusive ideological foundations, which are predominantly objective idealistic. To work out these different roots of the oppositional thinking of the time in individual cases, to determine the various traditional elements of thought, which the individual factions of the oppositional camp use for articulation, according to their origin and character—that is an extraordinarily complicated task.[47] But the most important reason for the contradictory form of oppositional thought becomes clear at the next stage of development.

2.3.4.3 Absolutism and 'Double Truth'

The first great upswing of the bourgeois mode of production simultaneously disintegrated the feudal system, in which the contradictory forces of feudal society had found a balance that was successful, at least in general; the civil wars began, in which the absolutist nation states emerged. Just as the urban bourgeoisie, in the struggle for urban autonomy, supported royal power against the respective local

---

all through P.J. Olivi and Franciscus de Marchia; cf. K.Esser, Anfänge und ursprüngliche Zielsetzungen des Ordens der Minderbrüder, Leiden 1966; on the civic orientation of these Franciscans cf. J. Kirchner, From usury to public finance—The ecclesiastical controversy over the public debts of Florence, Genova and Venice (1300–1500), Columbia U.P., 1970).—What remains open in all this is whether early nominalism not only produced ideas that could be used by the bourgeois opposition, or whether it also stemmed from this opposition. Even if there is hardly any evidence for the second, I think two considerations speak in favour of it. One is the question of why, from the late eleventh century onwards, the question of universals became the focus of philosophical debate in the areas of the highly developed urban bourgeoisie, although its main philosophical arguments had been familiar for centuries (both from Porphyry's introduction to Aristotle's *Categories* and from texts by Boethius) without arousing any particular interest. The other consideration is that an intellectually stringent connection can be shown between nominalist philosophy and the anti-clerical doctrines (especially in the consequence of the nominalist theory for the conception of the sacraments, on which the institutional role of the church, and thus its 'earthly' form, depended).—For the developed nominalism of the second half of the thirteenth century and for that of the fourteenth century, the connection with civic opposition is more significant. Siger of Brabant and above all Boethius of Dacia undoubtedly form a summary intellectual centre of all anti-feudalist opposition of the thirteenth century (cf. Gautier and Jolif 1958; Le Goff 1964; and H. Ley's modifying remarks on it, 1966, vol. 2/2, 160 f).—For the so-called terministic nominalism (Ockham, Buridan), the connection with the anticlerical enforcement attempts of the royal central power is familiar, although it can remain open here whether Ockham's intentions can be equated with the objective role he played for Louis the Bavarian (cf. Miethke 1959; McGrade 1974).

[47] Russel (1963).

nobility (and in the process often came under the rule of the king, who was primarily concerned with the money of the cities), so now, too, as it developed into the national class, it supported the absolutist form of the dictatorship of the large-landed estate, under which it could expect more appropriate conditions for development. It testifies to the sense of reality of the bourgeoisie when, in spite of all its imponderables, it recognized the absolutist rule of the land as the best possible condition, at any rate better than the previous condition, for the development of the bourgeois mode of production, and defended this advanced form of aristocratic rule against the more backward ones. This shows that the bourgeois mode of production needed the feudalist mode of production for its development, not only in its very beginnings, but even at this relatively advanced stage. The social context founded on the division of labour and the exchange of commodities proves not yet to be independent; it is not yet able to substitute itself for the old relations, but continues to have these—albeit in their absolutist form—as its prerequisite. Bourgeois society, though founded in itself, is nevertheless externally conditioned. The appearing social context is partial and presupposes another context, which was not to be understood as founded in the appearing reality.

Absolutism reveals in a sharpened form that bourgeois thought, where it did not surrender to the deceptions of limited local special conditions, was all this time only the mental representation of a contradictory social reality, whereby it is not a matter of the contradiction inherent in the bourgeois mode of production, but rather of the contradiction between it and the feudalistic mode of production that was still indispensable for it. Under these conditions, the social context that appeared in the bourgeois mode of production could be a support for oppositional thinking against objective idealism only in the form that the old dualism between the void, appearing world and the true, otherworldly world was turned upside down and the recognizable, true, appearing world was contrasted with an otherworldly world with a clearly irrational accent. This form of oppositional thinking can be traced in various forms from Latin Averroism and English Nominalism to Humanism and Protestantism.[48]

Nothing can be more nonsensical, in my opinion, than the reproach that thought thereby falls back behind the unity of earthly and divine "ordo" achieved in Thomism into the obscure dualism of Platonic-Augustinian character. On the contrary, through the break between the spheres, the new dualism creates free space for the interest in

[48] For the formation of Protestantism, the philosophy of Duns Scotus and the nominalism of Ockham are often attributed decisive importance, because in them the "deus absconditus" was preformed to a certain extent. However, this is at most justified with regard to the modified 'double truth' of Duns Scotus and Ockham, but by no means with regard to the concept of God (cf. Minges 1907). The 'double truth' seems to me to lie in the distinction made by Duns Scotus between theology and philosophy, which cannot be reconciled with the Thomistic interlocking of divine and earthly *ordo*, no matter how hard one tries to reduce the difference to a marginal one (cf. e.g. Harries 1959, 84 pass.). The difference appears—and this is of the greatest importance with regard to Protestantism—paradoxically that Duns Scotus does not draw any boundaries of knowledge for philosophy, because he because he assigns to it a territory sui generis (cf. Opus Oxeniense, Prol. q I, a I 6 b). This shows that Duns Scotus is by no means a forerunner of agnosticism or a pessimism of knowledge. or pessimism of knowledge. This is also true for Ockham (cf. note 50).

the appearing world, withdraws it from the fetter that the Thomistic interlocking of the spheres represented for a cognition directed at earthly things. The much-maligned "double truth" of Averroes proves to be the formula of the panier under which bourgeois thought asserted itself at a time when feudalism was still irrevocable. Only where at this time the bourgeois mode of production was able to develop under particularly favourable conditions to such an extent that it was able—as in northern Italy—to wrest an entire region from the feudalist mode of production, was it also possible to cast off this dualism. The northern Italy of the Renaissance, however, shows at the same time that this did not already eliminate the reasons that prescribed a speculative form for bourgeois thought.[49]

An elaborate account of feudal society, which between the thirteenth and seventeenth centuries was increasingly transformed by the bourgeois mode of production, but did not forfeit its essential features, would, as this rough sketch may perhaps make clear, make it possible to reconstruct essential parts of the contradictory currents in the genesis of modern science without any constraint on the material.[50] For the purposes of this work, however, it is important to see that the contradictions of the bourgeois mode of production are still hardly reflected in this contradictory

[49] In this context, in addition to Kristeller (1961); on Italy in the Renaissance, see particularly Tomberg (1973, 57–76).

[50] An instructive example of the violent misinterpretation of sources is offered by E. Conze (1932, § 3, 213 ff), who interprets Ockham's nominalism as subjective idealism.—This is in no way tenable. It is true that Ockham differs from the Northern French nominalism of the thirteenth century, which with increasing maturity took on an unequivocally materialistic character (cf. H. Ley 1966, vol. 2/2, para. 111), in that he ascribes reality to the general only as a concept of understanding (cf. Sum. tot. Log. I, C. XVI, in Ockham 1957, 37 ff). Nevertheless, it is self-evident to Ockham that the general concepts correspond to something in reality outside the mind, namely that the individual things do indeed have equalities. For him, therefore, the general concepts are not an arbitrary or transcendental, intersubjective or otherwise necessary product of understanding based solely on subjectivity, but rather processes of what the individual things—independent of our understanding—have in common objectively (cf. in particular the Commentary on Sentences I, d 2, q 4, M to 0 and ibid. d 27, q 3 AA; furthermore Exposition super librum Perihermeneias, in Ockham 1957, 43 ff, esp. 45; – cf. on this Hochstetter 1927, p. 178; Miethke 1959, 197 f). Likewise, it cannot be overlooked that Ockham adheres epistemologically to the materialistic conception of Aristotle (especially in the Second Analytics) (Cf. e.g. Quodlibeta I Q XIII, in: Ockham 1957, 27 ff. and Ordinatio D II, Q VIII (first version), Ockham 1957, 41 ff.; cf. on this Miethke 1969, 195 f.). This epistemological position, together with his position on the question of universals, precludes his introducing a separate mode of cognition for the universals to which he assigns a priority of *adaequence* (Quodlibeta, Ockham 1957, 32), as he would have to do as a subjective idealist—and as Themas did (cf. Hirschberger 1969, 66 ff.), who would therefore rather be claimed as Ockham for subjective idealism, namely precisely when one interprets, as usual, his so-called moderate realism as only differentiated from extreme nominalism, for whom universals are only flatus vocis. In Ockham's case, therefore, we have before us a kind of materialism that in many respects may already be reminiscent of later sensualism, but by no means a variety of subjective idealism.—Conze (1932) is not alone in his interpretation (cf. of more recent works e.g. Haag 1960, 21 ff. or Blumenberg 1965, 37–39). Conze's work, however, is of particular interest in our context because, in attempting to ground his interpretation in social history, he makes clear the relevance of a correct definition of the early bourgeois period under prevailing feudalism: He assumes that capitalist production structures had already developed in England at that time.

prehistory of modern science. Or more precisely: that the contradictory relationship of the bourgeois to the feudalist mode of production, in which each excludes the other and—still at the time of absolutism—at the same time demands it, covers up the specific contradictions of the bourgeois mode of production. The opposition to the world view of feudalism can, however, for this reason easily misjudge the contradictions in the social context that has arisen with the bourgeois mode of production, since these contradictions are undeveloped in the pre-capitalist stage and, as a result, are as yet hardly recognizable as specific contradictions of the bourgeois mode of production. Now it does not follow from this that at that time there were no effects at all of the contradiction of the bourgeois mode of production on thought; but it does follow that only the specifically historical form of the contradiction in the period under investigation here can provide insight into these effects.

2.3.5 *'Bourgeois Thought' in the Pre-capitalist Stage*

Until the sixteenth/seventeenth century, commodity production and its social contexts did not displace the preceding mode of production, but at best undermined it. In the process, the bourgeois mode of production imposes its stamp on the feudalist one. Just as the achieved unfolding of the bourgeois mode of production made possible the emergence of the absolutist form of the dictatorship of large-landed property, thus proving that it had advanced to become a constitutive moment of society, so it also coloured the productive life of society where it had not yet been able to assert itself. Even if the production of use values continued to predominate—apart from a few bourgeois islands such as northern Italy and Brabant—there is hardly a producer left in western and central Europe who did not also produce for barter or exchange commodities on the market. Apart from the specialization of whole agricultural regions on certain products, which had already begun earlier due to the sale of taxes in kind by the feudal lords, the transformation of the taxes in kind of the serf peasants into money taxes, which took place since the end of the thirteenth century, is probably the decisive factor in this connection.[51] The interconnection of private producers based on the division of labour across the borders of the nation states, which were only just being formed, thus tended to intervene in the lives of everyone.

The contradiction that this social whole, which abolished the lack of social coherence in feudal society, was for its part based on private labour, that is, on social production, but on production that was not carried out socially, has had the effect from the beginning, and so also for the period under consideration here, that the social whole between the private producers is realized only in the exchange of their labour results, their social relations appearing as the relations of their products. It is true that the developing division of labour, which no longer characterized the productive life of the cities alone, but also determined agrarian production, and thus

[51] Cf. Pirenne (1971, 84 f.); Kulischer (1958 vol. 1, 114 ff. and 255 ff.).

transformed it in the first place from an all-encompassing into a specific, precisely agrarian production, brought about functional dependence of the private producers upon one another, and thus produced a real, necessary social whole between them; but as private producers they could carry out this real connection only as buyers and sellers, i.e., as agents of the connection which existed among their products as commodities. The social whole of private producers that has come into being is determined and regulated in its execution by the value character of the products of labour, which is inherent in them as the results of private labour to be exchanged.

To conclude from this that a thinking which represents the social whole that has arisen with the bourgeois mode of production must reproduce it in the form determined by value, that this thinking must therefore contain the specific formal and abstract characters of exchange value analysed by Marx, seems to me extremely premature.[52] In general, it is to be objected that this form does not determine thinking as it is, but as it appears, and that therefore conclusions can only be drawn about thinking from the appearance of this form. Without this difference, political economy would be a superfluous science. In particular, however, the question must be raised as to whether, in the state of development of commodity production at that time, the appearance of commodity exchange did not have to contradict its lawful form in principle, so that this lawful form could not even be reflected in thinking in a distorted way.

At the time under consideration here, for the predominant part of the producers the exchange-value property distinguished precisely not all, but only particular products of labour, since these producers both produced only in part for the market and acquired only certain articles of need through the market. This applies not only to the countryside, but also to a large part of the cities. The value character, then, is for the greater part of the producers not the general form of the products of labour, but the special form of very definite products of labour. Under these circumstances it seems questionable how these producers could have comprehended the abstraction from the sensuous nature of the products of labour which is contained in the value character: only products with a certain sensuous quality seemed to have a value character at this stage of commodity circulation; value character and certain sensuous quality had to be regarded as coinciding.

2.3.6 *Self-Regulatory Social Context and Materialism*

However, even where commodity production and commodity exchange were most developed, they took place under conditions in which, on average, the lawful form of commodity exchange could not appear. The decisive condition, the medium in which the circulation of commodities can develop according to its nature, was lacking; in other words, unhindered competition was missing. Concessions and

[52] Cf. Sohn-Rethel (1978).

privileges, the most diverse customs duties at city and country borders, guild rules, the chaos of taxation, special levies, confiscations, and other violent expropriations weighed heavily on the circulation of goods and did not permit it to develop the form appropriate to it. The bourgeoisie, of course, was not united in its attitude to these restrictions, but was divided into the part which accepted them because it profited from them, and the part which saw its existence threatened by this medieval rubbish. In the struggle which this part of the bourgeoisie waged against all these abuses, including the practices of the other part of the bourgeoisie, there takes place, as it were, the epiphany of the regular circulation of goods, namely, in the fact that in this struggle all those special regulations are denounced as irregularities and disturbances of the circulation of goods, and their removal is demanded not for an already existing regular circulation of goods, but for the emergence of a regular one. The fact that this struggle—as English history shows—continues even after the revolutionary elimination of absolutism, namely in the mercantile age, incidentally draws attention to the fact that commodity traffic could only find its developed form when its basis, commodity production, had become dominant in all sectors.

The mental representation of social reality, therefore, could not in any way contain the social context that had arisen in the specifics of form that it receives through the law of value when commodity traffic has unfolded. In the critique of the disturbances and stagnations that commodity traffic suffered under absolutist conditions, however, it becomes apparent how thought represents the quid pro quo that the connection of the members of society has its appearance in the connection of their labour products. Criticism opposed arbitrariness and the manifold interventions and manipulations with respect for the inherent lawfulness of the social relations founded on commodity production, because this is to serve the benefit of all most easily and undisturbed; i.e., in this critique the social context, which arose with the bourgeois mode of production and which must be helped to its full breakthrough, is conceived as a relation that reproduces itself according to its own immanent and necessary laws, in the working of which it is not possible to intervene without negative consequences. This mental representation of the social context that has come into being with the bourgeois mode of production is incompatible with an objective idealistic worldview; it can only serve as a basis for a materialistic worldview, the ambivalence of which can be indicated here—it will be discussed in the fourth chapter—by the question of whether one must—with F. Bacon—"obey" the laws of this immanent, necessary context "in order to master them", or obey them because one cannot master them.

This critique and the conception of social coherence it contains is, of course, only consistently articulated between the end of the seventeenth and the middle of the eighteenth century (J. Locke and A. Smith may be taken as delimiting this period), at a time, that is, when the circulation of commodities and, above all, its basis, commodity production, were so far developed, at least in advanced countries like England, that one no longer thought of venturing out into chaos by entering the realm of unrestrained competition. But this does not affect our argument. For it follows, on the one hand, that even at the highly developed stage of the bourgeois mode of production immediately on the threshold of the industrial age, the mental representation of the contradiction of this mode of production consists in the fact that society is conceived as a self-regulating, natural connection. Whoever regards this view not as a mere apologetic manoeuvre to cover up the contradictions of the bourgeois mode of production, but, with Marx, as the expression of a "necessary appearance", is, in my opinion, compelled to see in the contradictions of bourgeois thought at still earlier stages of development the precipitation of social contradictions not on account of the developed, but on account of the undeveloped bourgeois mode of production, i.e. of contradictions which result from the fact that the bourgeois mode of production cannot yet develop according to its own laws.[53]

On the other hand, however, the development of the conception that in the bourgeois intercourse of the members of society there is a natural connection whose autonomy is to be respected can be traced back to the thirteenth century; this conception is thus by no means a fruit of the successful bourgeois revolution of the seventeenth century, even if its early forms naturally demand and assert the autonomous social whole more than they state it.[54]

2.3.7 The Contradiction: Speculative Materialism

As long as the bourgeois mode of production was not so far developed as to be able to reproduce itself as a system according to its own laws, but its character as a connection of immanent necessity became apparent only in individual phenomena, as long the materialist world view based on it had to be two-sided and contradictory and also had to express in some form the reality of feudal society, which was still a condition of reproduction of the bourgeois mode of production.

Under these conditions, the social coherence established by this mode of production was still to be asserted as a reality founded in itself; it was still the goal of the struggle of the bourgeois class and reached a realisation appropriate to it only in the minds of the strugglers. Thus, the materialistic world-view although a mental representation of this social reality, could not prove its truth by showing this reality as an immanent lawful coherence.

If this was not possible at the stage of development reached by bourgeois society, for it represented such a connection only in rudiments, this means that this social reality could be processed by the head only speculatively into a materialistic world picture. But herein lies a defect of the materialistic world-view at this stage, faced with which all its other half-measures seem almost beside the point. For the

---

[53] This sheds light, for example, on Hobbes' "bellum omnium contra omnes"; it is precisely not an adequate reflection of the developed bourgeois society, but of its undeveloped state, not yet finally emancipated from the old modes of production. How much more so for Grotius' natural law or Machiavelli's theory of politics and history!

[54] Cf. for the thirteenth century above all the criticism of the urban bourgeoisie of France of Philip the Fair's coinage practice, for the fourteenth century e.g. Nicole Oresme's expert activities on the reform of the coinage system. Cf. Stanka (1957, 187 ff), Estrup (1966, 97–116).

speculative form fundamentally contradicts the materialistic character of the world-view, since a materialistic world-view, unlike an objective idealistic one, cannot appear as a true knowledge of reality without passing the test and proving this truth on and in the appearing reality. Only the appearing reality can serve as a witness to its truth.

2.3.8 The Opportunity: Natural Science

It only becomes fully comprehensible how much the ideological confrontation of the bourgeoisie with feudalism depended on the image of nature when faced with this contradiction. The possibility and truth of the 'natural' social coherence of a bourgeois world, which was to be striven for and only postulated, and which was to be opposed to the one determined by arbitrariness under the dictatorship of the large landed property, had to appear as substantiated, even almost as proven, if it were possible not only to assert, but to be able to prove, nature as a coherence founded in itself and reproducing itself on the basis of its own laws, i.e. as a coherence of immanent necessity.[55] Thus the bourgeois conception of the world would have broken out of the speculative form that contradicted it; it would no longer be opposed to the objective idealistic conception of the world merely as another speculation, but would have been transformed into a provable knowledge attested by the appearing reality, and thus the feudalistic conception of the world into a mere assertion that is refuted by the appearing reality. Under these conditions, the elaboration of a view of nature based on scientific knowledge represented the only chance for the bourgeois worldview to shed the speculative form that contradicted it.

At the same time, we have here the answer to the question left open in the first chapter, whether and to what extent the development of the bourgeois mode of production demands the modern natural science: it imperiously demands it. However, not out of technical needs of material production, but out of needs of the class struggle to enforce the bourgeois mode of production (thus at least before the eighteenth century). Moreover, it becomes apparent that the mental representation of the social context that emerged with the bourgeois mode of production could not have been supported by another speculation on nature, but only by natural science. Thus, the bourgeois world view did not lead to a merely different "ideological" view of nature, so that modern natural science could not actually be regarded as science. It

[55] This and the following sentences are not to be understood as if some strategic head of the bourgeois opposition had made these considerations and thereupon recommended concentration on the natural sciences: The bourgeois class—like all classes before and after it in its ideological conflict—has directed its interest to the whole of reality—and not predominantly to nature or society—in order to secure its ideological position and to destroy that of the opposing class; no special recommendation was needed for this (cf. p. 86 f. **this is WL referring to German original of his book p 86**). The point here is to show that, under the conditions of the time, the bourgeois worldview had the objective possibility of verifying itself solely with regard to nature, although it was formed as a mental processing of the newly created social reality.

could only be regarded as the implementation of a certain "ideological" view of nature, which requires a view of nature founded on the knowledge of the objective regularity of nature, on natural science.[56]

The necessity and the possibility of a scientifically founded image of nature are, of course, two different things. An essential reason for the speculative conception of nature of the feudalistic epoch consisted, as has been shown, in the undeveloped state of material production, which in turn was founded in the state of development of the productive forces. Without further development of material production, there was no prospect of being able to place the image of nature on a scientific basis. The overcoming of the limitations inherent in the artisanal form of production at some, at first exceptional, points of material production, which can be observed from the late fourteenth century onwards, created the preconditions for the development of scientific knowledge of nature. It took almost 200 years, of course, for the study of the problems confronting practice in these exceptional sectors of production to lead to a state of knowledge on the basis of which natural theory could overcome its speculative form. Following on from the beginning of this chapter, it should be noted that Copernicus precedes this turning point, while the overcoming of the speculative form of natural theory makes its appearance in the work of Galileo and Kepler. The leap that characterizes the connection between Copernicus and Galileo/ Kepler was based on the fact that the knowledge acquired on the problems of advanced production was not developed enough to serve as a basis for a scientific theory of nature until the second half of the sixteenth century. Before that, however, the speculations on nature of the feudalistic world view could not alone be opposed essentially only by other speculations on nature; before that, not even a clear idea could emerge of what exactly it is that distinguishes a scientific from a speculative theory of nature. The bourgeois conception of the world demanded natural science, without it having been possible to determine yet exactly what science is. How the concept of science was formed in dealing with the practical problems of those advanced sectors of production is the subject of the following chapter.

References

Asser, G. 1965. Zum Verhältnis von Mathematik und objektiver Realität. *Deutsche Zeitschrift für Philosophie* 13 (2): 173–176.

Bensing, M., and B. Rüdiger. 1973. Einleitung. In *Politische Schriften, Manifeste, Briefe*, ed. T. Müntzer, 1524–1525. Leipzig: Bibliographisches Institut.

Blumenberg, H. 1965. *Die kopernikanische Wende*. Frankfurt am Main: Suhrkamp.

Blumenberg, H. 1985. *The legitimacy of the modern age*, trans. R. Wallace. Cambridge, MA: MIT Press.

[56] Borkenau, op. cit., 15–22, esp. 21 and 74; Marcuse (1965, 127). A constructivism operating with "protophysics" (cf. Lorenzen 1964, 143–150, translation Lorenzen 1987) is probably also to be placed here.

Borkenau, F. 1934. *Der Übergang vom feudalen zum bürgerlichen Weltbild*, Schriften des Instituts für Sozialforschung. Vol. 4. Paris: Alcan. (Repr. Darmstadt: WBG 1971).

Butterfield, H. 1949. *The origins of modern science*. London: Bell & sons.

Cassirer, E. 1932. Die Antike und die Entstehung der exakten Wissenschaft. *Die Antike. Zeitschrift für Kunst und Kultur des klassischen Altertums* 8:276–300. Die Antike XIII.

Cassirer, E. 1940. Mathematische Mystik und moderne Naturwissenschaft. *Lychnos* 5:248–265.

Chenu, M. D. 1969. Die Platonismen des 12. Jahrhunderts. In *Platonismus in der Philosophie des Mittelalters*, ed. W. Beierwaltes, 268–316. Darmstadt: WBG.

Conze, E. 1932. *Der Satz vom Widerspruch*. Frankfurt am Main: Verlag Neue Kritik.

Crombie, A. C. 1952. *Augustine to Galileo: The history of science A.D. 400-1650*. London: Falcon Educational Book.

de Gandillac, M. 1941. *La Philosophie de Nicolas de Cues*. Paris: Aubier.

Dijksterhuis, E. J. 1956. *Die Mechanisierung des Weltbildes*. Berlin: Springer.

Düring, I. 1966. *Aristoteles*. Heidelberg: C. Winter.

Engels, F. 1962. Feuerbach und der Ausgang der klassischen deutschen Philosophie. In *Marx Engels Werke (MEW)*, vol. 21. Berlin: Dietz Verlag.

Engels, F. 1990. Ludwig Feuerbach and the end of classical German philosophy. In *Marx Engels collected works*, vol. 26, 353–398. London: Lawrence & Wishart.

Esser, K. 1966. *Anfänge und ursprüngliche Zielsetzungen des Ordens der Minderbrüder*. Leiden: Brill Archive.

Estrup, H. 1966. Oresme and monetary theory. *Scandinavian Economic History Review* 14 (2): 97–116.

Freiesleben, H. Ch. 1969. *Galileo Galilei. Physik und Glaube an der Wende zur Neuzeit*. Stuttgart: Wissenschaftliche Verlagsgesellschaft.

Galilei, G. 1907. *Le opere di Galileo Galilei: edizione nazionale sotto gli auspici di sua maestà il re d'Italia*. Firenze: Barbèra.

Gammersbach, S. 1959. *Gilbert von Poitiers und seine Prozesse im Urteil der Zeitgenossen*, Neue münstersche Beiträge zur Geschichtsforschung. Vol. 5. Münster: Böhlau.

Gautier, R. A. and J. Y. Jolif. 1958. *Einleitung zur Nikomachischen Ethik des Aristoteles*. Paris.

Grabmann, M. 1948. *Die theologische Erkenntnis-und Einleitungslehre des hl. Thomas von Aquin auf Grund seiner Schrift. 'Boethium de Trinitate' im Zusammenhang der Scholastik des 13. und beginnenden 14. Jahrhundert dargestellt*, Thomistische Studien. Vol. 4. Fribourg, Switzerland: Paulusverl.

Haag, K. H. 1960. *Kritik der neueren Ontologie*. Stuttgart: Kohlhammer.

Harries, C. R. S. 1959. *Duns Scotus*. New York: Humanities Press.

Hegel, G. W. F., 1970. *Enzyklopädie der philosophischen Wissenschaften*, ed. E. Moldenhauer & K. M. Michel. Frankfurt am Main: Suhrkamp. Just cite in Werke?

Hegel, G. W. F. 1970–1979. *Werke in zwanzig Bänden, Theorie Werkausgabe*, ed. E. Moldenhauer & K. M. Michel. Frankfurt am Main: Suhrkamp.

Hegel, G. W. F. 1991. *Elements of the philosophy of right*, trans. H. B. Nisbet, ed. A. W. Wood. Cambridge: Cambridge University Press.

Hegel, G. W. F. 2010. *Encyclopaedia of the philosophical sciences in basic outline, Part 1, Science of logic*, eds. and trans. K. Brinkmann, and D. Dahlstrom. Cambridge: Cambridge University Press.

Hirschberger, J. 1969. Platonismus im Mittelalter. In *Platonismus in der Philosophie des Mittelalters*, ed. W. Beierwaltes, 56–72. Darmstadt: WBG.

Hochstetter, E. 1927. *Studien zur Metaphysik und Erkenntnistheorie Wilhelm von Ockhams*. Berlin-Leipzig.

Kalivoda, R. 1966. Zur Genesis der natürlichen Naturphilosophie im Mittelalter. In *Atti del III. Congresso internazionale di filosofia medioevale: La filosofia della natura nel Medioevo*, 397–406. Passo della Mendola: Vita e pensiero.

Kauffeldt, A. 1954. *Nikolaus Kopernikus - Der Umsturz des mittelalterlichen Weltbildes*. Berlin: Verlag Neues Leben.

Kirshner, J. 1970. *From usury to public finance: The ecclesiastical controversy over the public debts of Florence, Genoa and Venice (1300-1500)*. New York: Columbia University Press.

Klemm, F. 1954. *Technik - Eine Geschichte ihrer Probleme*. Freiburg: Karl Alber.

Koch, J. 1969. Augustinischer und Dionysischer Neuplatonismus. In *Platonismus in der Philosophie des Mittelalters*, ed. W. Beierwaltes, 316–342. Darmstadt: WBG. Wissenschaftliche Buchgesellschaft.

Kosminski, J. A., and S. D. Skaskin, eds. 1958. *Geschichte des Mittelalters*. Berlin: VEB Verlag der Wissenschaft.

Kosminski, J. A. et al. 1958. *Geschichte des Mittelalaters*. Berlin.

Koyré, A. 1969. Sciences exactes. In *Histoire générale des Sciences*, ed. R. Taton, vol. II, 12–207. Paris: PUF.

Kristeller, P. O. 1961. *Renaissance thought. The classic, scholastic, and humanist strains*. New York: Harper Torchbooks.

Kuczynski, J. 1961. *Die Geschichte der Lage der Arbeiter unter dem Kapitalismus*. Vol. 40. Berlin: Akademie-Verlag.

Kuczynski, J. 1972. *Wissenschaft und Gesellschaft: Studien und Essays über sechs Jahrtausende*. Berlin: Akademie-Verlag.

Kuhn, T. S. 1957. *The Copernican revolution: Planetary astronomy in the development of Western thought*. Cambridge, MA: Harvard University Press.

Kulischer, J. 1958. *Allgemeine Wirtschaftsgeschichte des Mittelalters und der Neuzeit*. Vol. 2. Munich: Oldenbourg.

Le Fort, G. V. 1962. At the gates of heaven. In *The judgement of the sea: Four novellas*, ed. G. V. Le Fort. Chicago: Henry Regnery Company.

Le Goff, J. 1964. Quelle conscience l'université médiévale a-t-elle d'elle-même? In *Miscellanea Mediaevalia*, ed. P. Wilpert and W. P. Eckert, vol. 3, 15–29. Berlin: De Gruyter.

Lefèvre, W. 1971. *Zum historischen Charakter und zur historischen Funktion der Methode bürgerlicher Soziologie*. Frankfurt am Main: Suhrkamp.

Lenk, W. 1974. *Einleitung zu Dokumente aus dem deutschen Bauernkrieg*. Leipzig: Reclam.

Ley, H. 1966. *Geschichte der Aufklärung und des Atheismus*. Berlin: Deutscher Verlag der Wissenschaften.

Lorenzen, P. 1964. Wie ist die Objektivität der Physik möglich? In *Argumentationen. Festschrift für Josef König*, ed. H. Delius and G. Patzig, 143–150. Göttingen: Vandenhoeck.

Lorenzen, P. 1987. How is objectivity in physics possible? In *Constructive philosophy*, trans. K. R. Pavlovic, ed. P. Lorenzen, 231–237. Amherst: University of Massachusetts Press.

Lukács, G. 1980. *The ontology of social being 3: Labour*, trans. D. Fernbach. London: Merlin.

Marcuse, H. 1965. *Kultur und Gesellschaft*. Frankfurt am Main: Suhrkamp.

Marx, K. 2010. Outlines of the critique of political economy (Grundrisse). In *MECW*, vol. 28. London: Lawrence & Wishart (Electric Book).

McGrade, A. S. 1974. *The political thought of William Ockham*. Cambridge: Cambridge University Press.

Mielke, H. 1966. Erkenntnisse und Irrtümer in der griechischen Naturphilosophie. In *Wissenschaft und Weltanschauung in der Antike*, ed. G. Kröber. Berlin: Deutscher Verlag der Wissenschaften.

Miethke, H. 1959. *Ockhams Weg zur Sozialphilosophie*. Berlin.

Minges, P. 1907. *Der Gottesbegriff des Duns Scotus*. Wien: Von Mayer and Co.

Mitteis, H. 1958. *Lehnrecht und Staatsgewalt*. Weimar: Böhlau.

Ockham, W. 1957. *Selected philosophical writings*. Edinburgh: Nemson.

Pirenne, H. 1963. *Histoire économique et sociale du Moyen Age*, ed. H. Van Werveke, Revised and expanded edition. Paris: PUF.

Pirenne, H. 1971. *Sozial- und Wirtschaftsgeschichte Europas im Mittelalter*. Munich: Francke Verlag.

Reidemeister, K. 1949. *Das exakte Denken der Griechen*. Hamburg: Claassen & Goverts.

Rietzler, K. 1940. *Physics and reality - Lectures of Aristotle on modern physics*. New Haven: Yale University Press.

Russel, J. 1963. Some interpretations of the origins of mediaeval heresy. *Mediaeval Studies* XXV:26–53.

Santeler, J. 1939. *Der Platonismus in der Erkenntnislehre des hl. Thomas von Aquin*. Innsbruck-Leipzig: Rauch.

Schenk, G. 1973. *Zur Geschichte der logischen* Form I. Berlin.

Sellnow, I. 1961. *Grundprinzipien einer Periodisierung der Urgeschichte*. Berlin: Akademie Verlag.

Sidorova, N. A. 1959. Петр Абеляр — представитель средневекового свободомыслия. In *Петр Абеляр. История моих бедствий (lat. Historia Calamitatum)*. Moscow: Akademiâ nauk SSSR. Otdelenie literatury i âzyka.

Smirin, M. M. 1952. *Die Volksreformation des Thomas Müntzer und der große Bauernkrieg*. Berlin: Dietz Verlag.

Sohn-Rethel, A. 1978. *Intellectual and manual labour: A critique of epistemology*, trans. M. Sohn-Rethel. Atlantic Highlands: Humanities Press.

Stanka, R. 1957. *Die politische Philosophie des Mittelalters*. Wien: Sexl.

Tomberg, F. 1973. *Bürgerliche Wissenschaft*. Frankfurt am Main: Fischer Taschenbuch.

von Schwerin, C. 1944. *Germanische Rechtsgeschichte*. Berlin: Junker und Dünnhaupt.

Wieland, W. 1962. *Die aristotelische Physik*. Göttingen: Vandenhoeck & Ruprecht.

Wohlwill, E. 1969. *Galilei und sein Kampf um die kopernikanische Lehre*. Wiesbaden: Sändig.

Chapter 3
The Significance of Practice for the Theoretical Form of Modern Natural Science

3.1 First Section

3.1.1 Interim Result

The first chapter of this book tried to show that, and to what extent, the development of the bourgeois mode of production simultaneously produced the preconditions for the genesis of modern science. It came to the conclusion that while on the one hand the bourgeois mode of production in fact produced materials, instruments, stimuli, fields of study, and even, to a certain extent, the individuals necessary for the development of modern science, on the other hand, it was not until the end of the seventeenth century that problems arose in material production whose solution required that material production be placed on a scientific basis. The conclusion drawn from this state of affairs was that the development of the bourgeois mode of production can be seen as a condition of possibility for the genesis of modern science, but that this genesis is not necessary from the point of view of the technical aspect of the material form of production specific to the bourgeois mode.

The second chapter investigated whether the necessity of the genesis of modern science can be found in the ideological position that the bourgeois class, supported by the new social relations it establishes, adopts and elaborates in its struggle against feudalism. The section concluded that the speculative form that the bourgeois worldview took in this stage of its development was in contradiction with the core of this worldview, which was ultimately materialistic, despite all its contradictions and half-baked ideas. This was due not least to the fact that the bourgeois worldview could not verify its assertion of an immanent and necessary coherence of appearing reality by invoking nature. Chap. 2 concluded by saying that the genesis of modern science became necessary in order to overcome the speculative form of the bourgeois worldview, which contradicted its content. At the end of the this chapter, I argued that the bourgeois theory of nature succeeded in overcoming its speculative

W. Lefèvre, *Theory of Nature and Mode of Production*, Political Epistemology,
https://doi.org/10.1007/978-3-032-02295-0_3

form in dealing with problems of the advanced sectors of production, i.e. in contact with practice, thereby launching the scientific form of a theory of nature, basic features of which have since then constituted the scientific character of theory in general.

3.1.2 Contact Between Theory and Practice

To a certain extent, this assertion expresses only a general opinion. Despite various justifications, assessments in detail and differences in emphasis, there is agreement among researchers in the history of science that the development of modern thinking between the fourteenth and seventeenth centuries—i.e. the development that can be called the breakthrough of modern thinking from speculation to science—could not be understood without the close contact between theory and practice which can be observed at least since the late fourteenth century.[1] The practical orientation cannot escape even those who study only the writings of the pioneers of this development. These writings show that it was part of the self-image of these pioneers to see the value of their theoretical work essentially also in its practical significance; partly they attached visionary perspectives for a happy future of the human race to this practical dimension of theory; this is not only true for F. Bacon, but similar outlooks can be found in large numbers among the scholars of the fifteenth and sixteenth centuries. The seventeenth century, however, as the publications of the scientific societies founded in that century show, emphasized the practical usefulness of science in such a way that one can almost see in it a fad-like trait of the spirit of the age.[2]

So there is no arguing about the fact that the contact between theory and practice had significance for the genesis of modern science from the late fourteenth century on. It is more difficult to determine exactly what the significance of this contact was. The fact that until the late seventeenth century practice owed little to theory, that is, that the practical usefulness of theory before that was generally only an aspiration and a promise, indicates that the contact that arose between theory and practice initially documented only a change in the conception of the meaning of theoretical knowledge, a demarcation from the understanding of theory that prevailed in the epoch of the dictatorship of large estates.

[1] I refer to the arguments of H. Blumenberg (1985), which contradict my own, because his work is disseminated here as a more recent account. Apart from the fact that this purely intellectual-historical work does not seem to me to meet the very requirements it sets for itself (e.g. his interpretation of nominalism), I cannot follow his central thesis, which is not exactly new, either: that the practical orientation of the modern age resulted from the fact that humanity, in a second, more radical gnosis, lost a supposedly given certainty of cosmological order. Blumenberg obviously did not escape the dilemma of the 'secularisation' thesis, which he discusses at the beginning of the book.

[2] The legal protection of technical inventions that was introduced in England in 1624 shows, of course, what a "fad" it was.

3.1.3 Theory as Elevation

The well-known passage about arithmetic and geometry in Plato's *Republic* (VII, 524 d–527 c) can stand here exemplarily for the understanding of theory, against which modern thinking turned with its emphasis on the practical significance of science. In this passage, where the influence of Eleatic philosophy is unmistakable, Plato speaks contemptuously of the shopkeepers who do mathematics "for the sake of buying and selling" (VII, 525 c3) and not in order to rise above becoming (525 b 5), above appearing reality, and thus to prepare their souls for the inner intuition of truth (the practical benefit that the military arm of the state derives from arithmetic and geometry is a different matter for Plato). The Greek landowning class, which since Parmenides made the illusion and nullity of appearing reality the starting point of its worldview, contrasted the world of the "double-headed mortals," this "deaf and blind at the same time" drifting, "stupidly gawking, undiscerning crowd", with a world of true being, the knowledge of which is the imprimatur of the aristocrat, of the man distinguishing himself from the crowd.[3] Theoretical knowledge, then, is a means of ennobling the individuals of a class, a means by which they can rise above the imperfections that inevitably cling to them in the spell of appearing reality; knowledge of truth is elevation above appearing reality.

Christian philosophy, born of the religious movements of the oppressed classes and peoples of the Roman Empire, eliminated the all too obvious emphasis on the elitist trait of this theoretical conception, which emphasized the supremacy of a certain class; but it took over the dualism underlying it and intensified it by the dogma of original sin, which made true knowledge without revelation seem unthinkable.[4] Theoretical knowledge initially lost any value at all and even came under suspicion of stabbing the fight against the "paganism" of antiquity in the back. A man like Augustine, educated according to the standards of antiquity, resisted the worst anti-theoretical radicalism of early Christianity by recommending that revelation be consulted as a decisive authority in non-religious questions only when it was a matter of significance in relation to the worldview and when it could not be decided with earthly knowledge. But at the same time he emphasized the vanity of all earthly knowledge going beyond ordinary, everyday knowledge. Until the threshold of the high Middle Ages, theory was essentially tolerated only insofar as it seemed to provide additional plausibility to the truth of revelation—a propaedeutic plausibility through eclectic "insights" into "creation,"—that is, in the role of a

[3] Parmenides, Fragment No. 6 (Parmenides 1922; Kirk et al. 1983, 271).

[4] In fact, it was only a shift in this elitist trait; by devaluing theory and declaring revelation to be the excellent source of true knowledge, the elitist character had to show itself only in a strengthened, administrative way, if the interpretation of revelation was reserved for the ecclesiastical authorities. Of course, this shift had to have important consequences: The struggle for and against the monopolisation of truth by one state had to take place in this starting position as a struggle between knowledge from revelation and theory in general rather than as a struggle between different theories.

"mnemonic."[5] Under the first significant onslaught of the bourgeois worldview in the thirteenth century, Christian philosophy took a catch-all position in Thomism, which lifted theoretical insight out of this inferior role. It passed from the status of a tolerated and basically superfluous means to that of an important means for paving the way to revelation. More precisely: for securing revelation against the undermining work of "pagan" philosophies. If in antiquity—apart from the theorists of the bourgeoisie of antiquity—theoretical knowledge was an outstanding means of elevating the human being over the changeableness of appearing reality, in the feudalist West it is a means (never considered without mistrust and in any case propaedeutic) to lead the human being to revelation and with it to the path supposed to lead him from the vale of sin and lamentation of appearing reality into a perfect world beyond.[6] In both cases theoretical truth has a practical use not in view of appearing reality, but only for the purposes of elevation above it.

3.1.4 Practice-Oriented Theory

The conception of the meaning and function of theoretical knowledge was thus closely related to the ideological positions that the landowning class and the bourgeoisie took against each other in their class struggle. In this ideological confrontation, theoretical knowledge on the part of the bourgeoisie took on a different function. Although it served both the bourgeois and the feudalist side to substantiate a worldview position, its accent had to shift to the cognition of appearing reality, since it was directed by the bourgeoisie against the degradation of appearing reality by the feudalist worldview. Appearing reality could no longer be the object of theoretical study only insofar as this study led to elevation above it; it rather became the object of theory on which its truth had to be proven. Knowledge of appearing reality became the center of theory.

Concentrated on appearing reality, this theory remained for a long time ambivalent, for reasons discussed in the second section, and above all was not able for the time being to get rid of its contradictory speculative form. Perhaps it can be interpreted as an articulation of an insight into this fundamental deficiency of bourgeois theory at this stage, when the pioneers of modern thought emphasized the practical utility of theory at a time when there could in fact be little talk of such utility; as an articulation of an insight, or perhaps only a hunch, that this deficiency can be remedied when theory works its way to real usefulness for practice.

---

[5] The "etymologies" of Bishop Isidore of Seville (560-636) are an impressive example of the regression of the view of nature.

[6] Cf. Tomberg (1973, 44 ff, esp. 51f).

3.1.5 Scholars Willing and Able to Cooperate

The circumstances under which the practitioners, who needed knowledge for the solution of certain practical problems, which could not be gained from their wealth of experience, and who therefore looked around for theorists who could possibly help them, found these theorists, speak for the fact that these theorists were willing to cooperate thanks to their ideological position. Let us imagine the situation. The practitioners had to find scientists to whom the relevant knowledge had been handed down through their education and who were also willing and able to translate, convey, and even work on what was useful to the practitioners. In other words, the practitioners were looking for people who were familiar with both the scientific tradition and the problems of practice. It is obvious that such scientists could not be found in every university town at that time.[7] At many universities, the very parts of the scientific tradition that were of importance for the practitioners were only barely cultivated, if at all—just imagine what it must have meant that until the sixteenth century at most universities of the West, arithmetic was still performed using Roman numerals. The inability of the majority of scientists to help the practitioners will as a rule have been coupled with the idea that they were above such lowly problems; the self-image of official science, remote from practice and protected by the guardians of the feudalist worldview, certainly represented the greatest obstacle to any dialogue between theory and practice. Nothing demonstrates this more strikingly than the fact that the practitioners found the people they needed among physicians and astronomers. If these scholars not only procured from the craftsmen and artists what they needed for their science, but were willing and, on the basis of a not only subject-specific interest and appropriate education, also capable of addressing the problems of the practitioners, of familiarizing themselves with them and of endeavoring to contribute to their solution with the help of the existing scientific and mathematical theories, it can certainly not suffice to explain that the scholars, who were indeed not well paid, were able to obtain additional income in this way.[8] No more distasteful view would be conceivable than that the scientific achievements of the fifteenth/sixteenth century were due to the fact that at that time it became economically rewarding for scholars to exert themselves. The interest and willingness of certain academics to collaborate in solving practical problems cannot be explained without the contradiction (discussed in the second section) between the speculative form and the content of the bourgeois worldview, and the resulting need to seize every opportunity for a theory committed to appearing reality. The problems of practice offered these opportunities in abundance. It is not by chance that the universities of Bologna and above all of Padua were the centers of activity and training of this type of practical scholar; these universities not only had in their neighborhood the best opportunities for study of all the technical arts of the time—one need only

[7] Cf. Koyré (1969, 28 f).

[8] The administrator, economist and physician Copernicus, who entered the history of science as an astronomer, is by no means anything out of the ordinary in his universal literacy at this time.

think of the arsenal of Venice and the factories of Florence; it was at these universities, above all, that the most developed form of bourgeois materialism, which had emerged in the thirteenth century—heterodox Averroism—had survived persecution best and had continued to develop.[9]

The cooperation between theory and practice at this time is thus an expression of the transformation in the conception of the function of theory, founded in the ideological demarcation of the bourgeoisie, which is revealed in the concentration of theory on appearing reality. That the contact between theory and practice supported this change of function and strengthened it anew, and that it can thus be called in this respect a co-conditioning moment of this change of function, probably does not require explanation. But nothing is gained from this insight for the initial assertion that modern thinking achieved its scientific form through contact with practice. For the time being, it only seems plausible that contact with practitioners and their problems did promote the empirical attitude of theory, not only morally, so to speak, but above all by the factual conditions of an empirical theory provided by practice (instruments, materials, broadening of the horizon, etc.).

3.1.6 Empirical Attitude

The empirical attitude of theory is undoubtedly an essential moment of modern science; to a certain extent, it is even justified to say that the empirical attitude is the characteristic feature distinguishing modern science from the theory prevailing in antiquity and the Middle Ages. If, for example, the view of Aristotle that the speed at which bodies fall increases or decreases in direct proportion to their weights remained unshaken for 2000 years, although everyone could have been convinced of the inaccuracy of this view without great effort by trial and error, then this speaks in favor of seeing the immunity to experience as the characteristic feature by which the theory of antiquity and the Middle Ages differs from that of modern times.[10] As

[9] The prominent role of the universities of Bologna and Padua in the genesis of modern science is hardly controversial; the significance of Averroism and Alexandrianism is perhaps a different matter. Cf. on this Gilbert (1963); Kristeller (1961), as well as the same and Randall (1941); Ley (1966, vol. 2/2, 439 pass.) as well as the same, 1957; Maier (1966, vol 1, II 90); Kuksewicz (1965). We will return to this later (p. 107 f).

[10] Cf. Aristotle, *Physics* b. 8. Of course, Aristotle did not intend to establish a law of falling bodies, but came to speak of the falling motion and its moments in order to moments in order to refute the atomistic doctrine of the vacuum (cf. Wolff 1971, 11–22). The errors that contained in the provisions on the proportionality of the moments of the falling moments of the falling movement have been noted and partly corrected by many commentators (cf. and partly corrected (cf. ibid.)). However, this does not alter the fact that these corrections did not have the effect that from then on everyone had to read this Aristotle passage as erroneous; rather, this passage could always be used again as authoritative evidence for the determination of the falling motion. This example, then, signals not only immunity to experience but, moreover, disinterest in the reality that appears.—This, incidentally, throws light on the view that the tacit adoption of impetus physics by large sections of scholastic scholars (cf. Maier 1966, vol. 2, II 1 0), which could be observed in the six-

is well known, that various peripatetic astronomers refused to follow Galileo's invitation to look through the telescope in order to convince themselves of the correctness or incorrectness of his assertions about the moons of Jupiter and about mountains on the moon is not a merely decorative anecdote in the history of science, but is a testified fact. The fact that Galileo pointed the telescope at the sky and that the Peripatetics refused even—as Galileo reported—to look at the telescope itself clearly demonstrates the difference between these two distinct approaches to theory.[11]

This difference is also evident from the fact that the emphasis on experience was directly directed against the theoretical method, which decided a question by limiting itself to weighing and interpreting the known doctrines about it according to the reputation of the various experts. The meaningful sifting of existing doctrines on a question—the incorporation of which as an integral part of theoretical investigation is certainly to be counted among the merits of Aristotle—had degenerated in the theoretical enterprise, remote from reality under the dictatorship of the large landowners, to the fact that "facility to believe, impatience to doubt, temerity to answer, glory to know, doubt to contradict, end to gain, sloth to search, seeking things in words, resting in part of nature...and the like" characterized the theoretical procedure, a procedure that people like Bacon detested not least because they considered it incompatible with human self-respect.[12] Immunity to experience, by the way, was not the root of the depravity of traditional theory for Bacon, who is considered the philosopher of empirical science; he attributed this immunity, like all other infirmities, to the remoteness of this theory from practice.

3.1.7 The Other Bible

The empirical attitude of theory is, as we have seen, an essential moment of modern thought, because it was directed against the dualism of the feudalist worldview. The empirical attitude of theory is at the center of the process which raised appearing reality from an inferior rank to that of a source of truth equal to the Christian revelation; though revelation remained supreme for religious questions, for all others, appearing reality became the only authoritative source of information. Perhaps the most interesting document for this ideological thrust of experiential theory is

---

teenth century, makes Galileo's criticism of Aristotelian physics appear to be a case of knocking down open doors; Galileo was in fact attempting with his criticism to put an end to precisely this disinterested tolerance with regard to physical conditions.

[11] Letter to Kepler of 19. 8. 1610, cf. Blumenberg (1965, 7).

[12] Bacon, 'In Praise of Knowledge' (c. 1591–1592), in Bacon (2012, 265), cit. and discussed in Horkheimer and Adorno (1947, 13), Horkheimer and Adorno (2002, 1); cf. *Novum Organum* I, Aph. 71. Remarks such as this aphorism should be examined to see to what extent this critique of antiquity is also meant to apply to. Renaissance philosophy. This view is, of course, already implied in Plato; cf. *Theaetetus* 152 e, *Sophist* 241 d-251 a, etc.

Galileo's letter—indirectly intended for the eyes of the Roman inquisitors—to the Grand Duchess Christine of Lorraine (1615), because in it Galileo tried to convince the Church that it was best for the assertion of its authority to accept this division of tasks between theology and science.[13] In this letter Galileo developed the thoughts that he later summarized in the *Saggiatore* (1623), taking up the famous metaphor of the "Book of Nature", whose thrust—to place another Bible alongside the Bible—no contemporary could misunderstand. "Precisely this assertion," writes Ernst Schumacher, "that the human spirit has a direct share in the divine spirit via knowledge of nature, must have annoyed the theologians in several respects. On the one hand, they must have felt disadvantaged because their participation remained limited only to faith; on the other hand, they must have seen in such a view the old Protestant spirit which declared mediation by the church superfluous; thirdly, they had to see in Galileo's identification of knowledge of nature and truth a profanation of the concept of truth par excellence, from which a materialistic worldview had to emerge.".[14]

3.1.8 Critical Empiricism and Experiment

The empirical attitude is thus undoubtedly a constitutive component of modern science.[15] But it would be extremely misleading, as virtually all historians of science point out, to think that modern science prevailed against a type of theory that knew no empiricism at all.[16] Rather, it contrasts a superficial empiricism with a critical one. The cited reception of the Aristotelian view of fall velocity may stand as an example of the superficial empiricism that was unacceptable to modern science as a starting point for theory. However, modern science also turned against the other extreme of empiricism, against "that scholasticism, that sophistry of empiricism, which only burrows from the particular into the particular again and again, restlessly differentiates into the infinite, subtilizes and specifies, leads us astray, makes of nature a labyrinth without exit …"[17] Both extremes coincide in that it is an eclectic empiricism that uses empirical data only for illustration; empirical inquiry is not embedded in a systematic analysis of appearing reality, in which facts not only make a theory more plausible or more questionable, but support it or deprive it of ground.

How fundamentally the empiricism of modern science differs from this is evident in what can be regarded as the epitome of critical empiricism, the experiment,

[13] Cf. Galilei (1895) (vol. 5), 307–349 (*Lettera a Madama Cristina di Lorena*).

[14] Schumacher (1964, 104).

[15] This constitutive role of empiricism only fully emerges in connection with the importance of analysis; we will return to this in this section.

[16] Cf. e.g. Blumenberg (1965, 36).

[17] Feuerbach (1969, 55).

the idea of which can be traced back in the West to the thirteenth century (Roger Bacon), but which only at the end of the sixteenth century (with Galileo) becomes the genuinely practiced method of empirical investigation. In the experiment, critical empiricism is revealed as a methodical questioning of appearing reality. Apart from its technical preconditions, the methodical and systematic character of this empiricism is based on the fact that it serves a theory which contains an explanatory model of the phenomenon in question based on existing experience already before it begins to experimentally interrogate the phenomenon in question; that it anticipates the valence of the data to be collected for the explanation.[18] In experimental empiricism, then, the aspects of the phenomena to be investigated already have the character of elements of a theory; their empirical investigation is a test of whether or not they do justice to their anticipated role in the structure of the theory. In this respect, we can say that critical empiricism investigates phenomena only as the theoretical explanatory model underlying the investigation "constituted" them.[19]

3.1.9 *The 'Shallow Empiricism of the Practitioners'*

Modern science not only set itself in opposition to the eclectic empiricism inherent in the theory prevailing in antiquity and the Middle Ages with this systematic empiricism; it also distanced itself from the practitioners with whom it had learned to cooperate. Galileo, whose theoretical and practical occupation with problems of material production continued from his apprenticeship years with Ostilio Ricci to his collaboration in old age in the regulation of the Arno and Bisenzo; Galileo, who still opened the late *Discorsi* with a praising reference to the arsenal of Venice, thus about whose close connection to material production there can be no doubt, is in his

[18] Incidentally, this precondition of experience shows how unreasonable it would be to demand experimental empiricism even in the initial stages of an experiential science discipline. It requires extensive observations with at least problematic attempts at classification and systematisation so that an explanatory model can be formulated from which experimental investigations can be carried out.

[19] As is well known, idealistic conclusions have been drawn from this theoretical character of critical empiricism. Idealistic conclusions have been drawn, which, in the extreme, went so far as to fundamentally reject the empirical the empirical character of the theory that thus interrogates the phenomena. (This was especially the case in the wake of the so-called Heisenberg indeterminacy relations. Indeterminacy relations; but it seems to me that logical positivism and the so-called positivism and so-called constructivism also belong here to a certain extent. Belong here to a certain extent. Cf. e.g. the statements in the fourth and sixth parts of Krüger (Ed.) 1970; cf. also Hörz (1974), Chaps. 4 and 8); for our context this would have the consequence that empirical attitude could not be regarded as an essential feature of modern science. of modern science. This is not the place for a fundamental discussion of this view, and it must therefore suffice to point out that the explanatory model that 'constitutes' the phenomena under investigation in the sense described represents the processing of experience and is not immune to further experience. What the semblance of constitution of the empirical foundations of a theory leading to such conclusions is based on will be discussed at the end of this section.

well-known reserve towards practitioners probably the exemplary figure of this distance; his distant attitude was above all also directed at the "shallow empiricism of the practitioners" (G. Harig).[20]

If the eclectic empiricism of the theory prevailing in antiquity and the Middle Ages resulted from its distance from practice, the unsystematic, "shallow" empiricism of the practitioners, conversely, resulted from an all too immediate orientation to practice. For the shallow empiricism of the practitioners was undoubtedly due to the fact that they were under pressure to constantly produce usable results; this not only took away the time they had to study a matter thoroughly, but also created in them the disposition to be satisfied with case-by-case solutions as long as it was at all possible. Thus the practitioners were not superficial empiricists in the sense that they did not observe closely enough; but it was impossible on the basis of their casuistic disposition to develop the systematic empiricism that can serve as a basis for a scientific theory.[21]

3.1.10 Conditions of the Contradiction

What emerges here, however, is not an eternal contradiction between theory and practice, but a specific historical one. The undeveloped level of material production is reflected in the casuistic attitude to knowledge from which shallow empiricism results. Practitioners, as the first section showed, were not interested in scientific knowledge because there might have been a need in some sector of material production to put production on a scientific basis. There were no conditions for this, neither on the side of material production nor on the side of theory. The traditional form of providing the intellectual conditions of labor, the transmission of the empirical wealth of experience from generation to generation, be it through family ties, apprenticeship, or journeyman years in the guild system, still corresponded in general to the state of development of material production. Reserves of experience could even be set free if the local and regional restrictions on the acquisition of experience could be overcome and the scattered wealth of experience made generally accessible. Mobility and transport as a result of market expansion worked in this direction. The technical literature that emerged from the fifteenth century onwards can still be interpreted as the highest stage of development of this traditional form, in which material production assured itself of its intellectual conditions at the level of the time, even if this literature enabled and to a certain extent already marked the transition to the assimilation of scientific knowledge into intellectual working conditions. But even where the conventional basis of material production was transcended, where the social form of labor and the use of machines demanded

[20] Cf. Klemm (1954, 170 ff); Harig (1964, 98); cf. also Olschki (1919–1927, vol. 3, 238).

[21] Cf. comments on this by the often misunderstood "shallow" empiricist F. Bacon, e.g. *Nov. Org.* I, Aph. 99.

a change in the form of the intellectual moment of labor, the interest of practitioners was not yet directed towards scientific theories, as can be seen in the Italian city states of the fifteenth and sixteenth centuries.[22]

The war technology of these peaceless states, their building industry, the harbor and ship operations of Venice and Genoa, the development of machinery, which became economically urgent due to the growing shortage of labor, especially in the sixteenth century when the collapse of the trade supremacy of northern Italy already led to the return of parts of the urban population to the countryside,—all this did not make practitioners call for elaborate scientific theories, but for knowledge that directly contributed to solving certain practical problems, regardless of where this knowledge came from and how it came about. A glance at the life of the physician Cardano or the jurist Vieta shows how these men were approached by merchants and engineers for assistance; just as one goes to a lawyer not because one wants to know something about jurisprudence, but because one needs legal assistance in a matter, so engineers and merchants at that time turned to natural scientists and mathematicians not to learn about their sciences, but for scientific assistance in solving a practical problem. Under these circumstances, it is astonishing and admirable that a self-taught man like Tartaglia, who made his living by providing engineers and merchants with special mathematical formulas for their respective problems, wrote mathematical and physical treatises that are far beyond a casuistic question.[23] As little as the practitioners, who at that time tried to use knowledge that could be gained theoretically for the solution of practical problems, can be equated with craftsmen, they too are unmistakably afflicted with artisanal casuistry, i.e. an attitude to knowledge that is incompatible with systematic empiricism.[24]

The inevitable contradiction between theory and practice at that time perhaps becomes even clearer when we consider the abstract character of experimental empiricism. Experimental empiricism does not look at the phenomenon in its concrete richness, but separates the different aspects of the phenomenon, concentrates on certain of these aspects, and not only temporarily neglects others, but seeks to practically eliminate or dispose of them through the experimental order. It is precisely this abstract character of empiricism, without which systematic natural science is inconceivable, that became problematic, even intolerable, for practitioners at the very point where abstraction was to be made from aspects of the phenomenon from which the practitioner could not abstract without endangering the purpose of the work. While, for example, physicists had to abstract from friction in certain research on mechanics, a practitioner could not be expected to make this abstraction; what was the practitioner, who had to show practical results within certain time limits, to do with findings that were only "correct" when abstracting from friction,

---

[22] Cf. footnote 28 in Sect. 1.

[23] A very vivid description of this intercourse between practitioners and scholars is given by N. Tartaglia in his *Nova scientia* (1537); cf. Harig (1958, 439) and Olschki (1919–1927, vol. 3, 73 ff.).

[24] Cf. Klemm (1954, 154 ff.).

i.e., in the practical case of interest, not at all or only very approximately?[25] Didn't practitioners even have to suspect that scholars who abstracted in such an "unrealistic" way were again moving closer to unworldly, speculative theory?

Here, the historical nature of the contradiction between theory and practice becomes particularly clear. The question of abstraction is, after all, a thoroughly technical question; what role friction plays in certain cases of material production, for example, whether it is a negligible quantity or not, depends on such seemingly banal things as lubricants, greases, the design of bearings, etc. The artificial practice of experimentation, on a small scale and precisely as a form of non-serious production on the same technical basis, was able to perform more far-reaching practical abstractions (e.g. vacuum generation) than typical material production. Theory, moreover, was able to make intellectual abstractions that—based on such experimental abstractions—no longer had to be groundless. The contradiction between theory and practice is thus revealed here as a direct consequence of the state of development of the productive forces.

3.1.11 A Renewed Separation of Theory and Practice?

After all this, the contact between theory and practice seems to have a very contradictory significance for the development of modern science. On the one hand, it is an expression of the change in the conception of the meaning and function of theory that emerged in the ideological struggles between the bourgeoisie and the aristocracy; cooperation with practitioners reinforced the concentration of theory on appearing reality. On the other hand, the very step it had to take to attain the level of a systematic, scientific theory showed it to be at odds with and distant from practice. The assertion made at the end of the second section, that theory, in contact with practice, shed its speculative form and discovered its scientific one, seems untenable. On the contrary, theory had to separate itself from practitioners and their problems in order to develop and maintain its scientific form.

The scientific achievements of the Renaissance in particular seem to confirm this. The outstanding theoretical achievements of the 15th and at least the first half of the sixteenth century are—apart from the theory of the state (above all Macchiavelli)—conspicuously remote from practice. The theory of this epoch is shaped by the philosophy of Nicholas of Cusa, by humanism, by Copernican

[25] This only approximate validity of theoretically determined regularities is linked to the well-known 'idealisation' theorem, which elevates this difference to a supra-historical and principled one. According to this theorem, modern science does not seem to be the science of reality in the literal sense, but the science of an idealised reality. According to this, there is only a rough, approximate correspondence between the reality that appears and the ideal reality dealt with in science. As is well known, Husserl went so far as to state that in modern science there is not a tendency towards the approximation of the ideal reality to appearing reality, but rather the opposite, a tendency towards the approximation of appearing reality to the idealised reality; cf. Husserl (1992). The exhaustion method of geometry was regarded by him as an exemplary indication. We will return to this at the end of this section.

astronomy and by further developments in mathematics, i.e. by theoretical achievements whose practical relevance is at least extraordinarily mediated.[26] The emphasis of theory in this period is perhaps most clearly evident in its preoccupation with the legacy of antiquity. The appropriation of ancient literature, which is important for a theory concerned with technical problems, is by no means absent; the translation of the works of Archimedes and Heron of Alexandria into more accessible Latin, for example, falls into this period.[27] But the main emphasis of the inheritance of antiquity at this time is undoubtedly on objects such as literature, rhetoric and philosophy, in other words, on objects that relate primarily and in a very mediated way not to the practice of production but to political practice.[28]

This is not contradicted by the fact that Alberti's *Trattato della pittura* and *De arte edificatoria*, Agricola's *De re metallica*, the work of Paracelsus, Vesalius's *De humani corporis fabrica*, and above all Leonardo's work, works that stand for the close connection between theory and practice in this stage of the development of modern science, were written in this period. Although it is perhaps justified to see these works as the main achievement of the Renaissance in terms of further development, they cannot be regarded as theory compared with the Copernican system or the algebraic works of the time; they are rather attempts to systematize experience and knowledge gained in practice, some of which are highly developed (especially Vesalius), some pragmatic (Agricola) and some extraordinarily obscure (Paracelsus,

[26] F. Bacon was not the first to criticise the humanists' remoteness from practice; Leonardo Da Vinci attacked it (cf. the 'Codex Atlanticus', 117 r.b, in Da Vinci 1955, 57). That Copernican astronomy did not emerge from a preoccupation with problems of practice was shown in the first section. With regard to mathematics, differences have to be made for its individual fields. The development of the non-geometric determination of proportions (Leonardo, Dürer, Ries and others) is directly related to the needs of perspective painting. of perspective painting; the systematisation of the use of decimal fractions (Regiomontanus, Vieta, Stevin) or the development of the modern notation of notation (Stevin) is certainly related to the needs of bookkeeping —Stevin, as is well known, was an accountant, the development of logarithms (Neper or Napier and Bürgi) with the needs of accounting of accountancy, engineers and captains. Such connections are, however, irrelevant to the progress in the field of algebra, which is undoubtedly the outstanding mathematical achievement of the time (solutions of third-degree equations: del Ferro, Tartaglia, Cardano; of fourth-degree equations: L. Ferrari; establishment of the algebraic symbolic language: Vieta). It is true that practically mathematicians have also been occupied with mathematical questions that arose from technical problems. Mathematical questions that arose from technical problems; it can certainly also be said that the that the problems of kinematics posed by ballistics are the result problems of kinematics raised by ballistics urged a further development of algebra. is directly evident in Tartaglia's work, for example. But the solutions found are not directly connected with these problems. Their practical significance only became apparent later, e.g. in the works of Kepler and Newton, which would have been inconceivable without the mathematical advances of the Renaissance.

[27] Cf. Koyré (1969, 28n).

[28] At this time, however, Lucretius' *De rerum natura* is opened up (cf. Crombie 1952, 38). However, as important as this development was for the physics of the seventeenth century, it is obvious that for the time being only a speculative use of the atomistic materialism of antiquity was possible; and this, as is well known, not only until the seventeenth century.

but to a certain extent also Leonardo).[29] This is perhaps clearest from the fact that physics, which dominated later in the seventeenth century, at this time occupied at best an equal place among the scientific disciplines, if not to say that it was on the inferior level of chemistry, above which, for example, medicine had already risen.

In their difference in level to those genuinely theoretical works of the Renaissance, even the practice-oriented achievements thus seem to confirm that modern theory had to break away from its close ties to practice in order to become science, for example, in the sense of the Copernican system. Now, as discussed in the second section, the breakthrough of modern natural science was not about attaining a certain theoretical level, but primarily about overcoming the speculative form of theory, whereby, among other things, the speculative use of mathematics, which was characteristic of the Renaissance, had to be overcome. Distance from practice could not contribute anything to this, as the theory that prevailed in antiquity and the Middle Ages shows; for its speculative form was based not least on its remoteness from practice. If the distance from practice at the end of the sixteenth century, exemplified by Galileo, did not lead theory back into speculation—as it obviously did—then it must have been a fundamentally different distance from practice than that which had been characteristic of theory until then. Thus, it must be assumed that the modern theory of nature had already taken the decisive steps before this distance, i.e. during the fifteenth and sixteenth centuries, which protected it against any relapse into speculation.

These decisive steps were taken, as will be shown, in the attempt to apply parts of the traditional and further developed theory, namely mathematical theory, to the solution of practical problems within the limits set by the state of the productive forces. Theory was able to develop in this attempt because the contradictions that inevitably arose in the process gave expression not only to the undeveloped level of material production but also, and certainly more so, to the undeveloped level of theory. Theory was able to profit in these attempts to apply it despite the pre-industrial level of development of production. The development of theory in these attempts at application can only be traced here at a few points that can be regarded as essential from the point of view of the result; in the process, many problems that arose in the ideological conflict between the bourgeoisie and the aristocracy will be found as problems of inner-theoretical development.

3.2 Second Section

3.2.1 Lorini

The problem of the merely approximate validity of theoretical statements within appearing reality, in which the different possibilities of abstraction of theory and practice at that time were expressed, can serve us as a starting point in

[29] Dijksterhuis remarks (1956, 285): " … it is as if he (Leonardo) was only concerned with the rules with which he can solve concrete tasks … ". But cf. also Olschki (1919–1927, vol. 1, 253ff.); Koyré (1969, 88 ff); Feldhaus (1930, 390 et seq).

understanding the development of theory in question. Lorini, a Florentine engineer (born 1545), raises this problem so vividly that he should have his say here in some detail.

> I will now summarize, as simply and briefly as possible, the effects of the lever on pulleys, screws, the axle and the wheel. These findings contribute most to the understanding of what we have to say about the study and manufacture of machines and how these are not only to be assembled and arranged according to the correct proportions, but also how, with the help of the compass, the force, i.e. the increase of the leverage, can be found in an understandable way, so that then, in the actual execution of such works, one is not deceived about their efficiency, as often happens to those who rely solely on the ease with which small models run, without knowing about the necessary principles. But before we go any further, I must point out the difference that exists between a purely speculative mathematician and a practical mechanic. This difference is due to the fact that proofs and ratios derived from lines, surfaces and merely imaginary, immaterial bodies no longer apply accurately when applied to material objects, because the mental conceptions of the mathematician are not subject to those impediments which are inherent in the matter with which the mechanic works ... And therefore the insight of the mechanic, who should order and command those who execute a work, consists for the most part in foreseeing the difficulties which lie in the quite different properties of the substances with which one must work.... And one should be all the more cautious the less certain rules can be given with regard to such obstacles... .Therefore, by what I have said above, I want to make it clear to those who want to embark on such undertakings ...that they not only need knowledge of mathematics, but that they must also be a careful and skilled mechanic ...[30]

3.2.2 *Approximate Validity*

In the cooperation between practitioners and mathematical theorists, both sides thus experienced that a mathematically stringently derived result had only an approximate (and often no) validity for practice. It is, of course, no coincidence that this problem manifested itself primarily in the application of mathematics. Until the middle of the sixteenth century, the practitioners' demand for scientific support was predominantly directed towards mathematical help. Later, the same problems manifested themselves precisely in the mathematical form in which the physicists treated their objects. In the abstractions underlying mathematics, the differences between theory and practice with regard to the possibilities of abstraction naturally had to show themselves in a potentiated form. On the other hand, this abstraction could not be dispensed with; it was the basis of the stringency and lucidity of mathematics that made it the paradigm of true knowledge for philosophers from Plato to Descartes. The contradiction experienced in the application of mathematics seemed to push once again towards Platonic dualism, towards the solution of separating a world of pure forms from the world of appearances, which only contained them in a distorted form, and devoting oneself to the study of the former as the sole object of true

[30] Quoted in Klemm (1954, 157 ff.), which will resonate with Galileo's later formulation of the problem.

knowledge. If the theorists of the Renaissance were not dissuaded by this contradiction from attempting to use the possibilities that mathematics contained for the solution of practical problems belonging to appearing reality, and if in doing so they disregarded some time-honored principles of mathematics, then this once again throws light on the change in function of theory that has been discussed.

3.2.3 The Desecrated Euclid

In antiquity, a slave-holding aristocracy of landowners, which was therefore uninterested in technical problems, used mathematics to forge ideological weapons from its problems of abstraction against the materialistic currents of thought that had arisen with the flourishing of various bourgeois city states.[31] For this aristocracy, the purity of abstraction had to be the decisive factor in mathematics; mathematical abstractions had an intrinsic value for them and became independent ideal realities.[32] For the theorists of the Renaissance, on the other hand, who wanted to draw practical benefits from mathematics, the purity of abstraction could not be an end in

---

[31] As is well known, the members or partisans of the Greek aristocracy not only 'used' mathematics, especially geometry, for their ideological purposes, but they also earned the merit of having developed geometry into a science in the first place. The distinction made in this science between the imagined elements and forms of space, which alone can form the basis of scientific proof, and the forms of appearing reality was undoubtedly an important point of support for the dualistic world view, in order to declare the world of pure forms (of course not only geometric ones) to be the only true one and thus at the same time to denounce appearing reality as a world of mere appearance (Klaus and Kröber 1966, 109), without which geometry could not have reached the level of a systematic science. The contradictory significance of the Greek aristocracy's distance from material production cannot be ignored in this context. For practical purposes, geometry in the form of undifferentiated rules of procedure was sufficient at that time, as the Babyionians and Egyptians prove, who, because of their problems of irrigation and drainage, had to make much more extensive use of geometry than Greece (Graecia Magna), which was carefree in this respect. Possibly it was due to this constant practical compulsion to practise applied geometry that the priestly scholars of Egypt and Mesopotamia, in contrast to the theoretically active members of the Greek aristocracy, did not go beyond a—admittedly highly developed—casuistic form of geometry. There is perhaps no need for a debate about the 'cunning of reason' that can be discovered in this; the contradictions in the cooperation between theoreticians and practitioners at the beginning of the modern era also show how much there is to be said for the fact that people who have to solve practical problems would only have succeeded with difficulty, or perhaps not at all, in the "mathematical negation of reality" that was only possible for geometry in antiquity.

[32] This view was already noticeable in the early days of the Pythagorean school, that is, at a time when Greek mathematics was only at a crude preliminary stage of idealistic mathematics, which Plato, under the influence of the Eleatics, laid the actual foundation for and which Euclid completed. At this early stage, the mathematical form had not yet been worked out purely; the speculation that wanted to base the appearing world on numbers implied that the numbers were understood as material building blocks. But this materialistic-looking variant of Pythagoreanism, which attempted to interpret the numbers analogously to the material principles of the Ionian natural philosophers as principles of all that exists, makes the mathematical forms into intrinsically founded, independent entities.

itself; any reduction that helped in practice without endangering the achievement of the mathematical method had to seem permissible to them. Thus these theorists did not hesitate to interpret the line as a sum of points and thus the point as something extended, the surface as a sum of lines and thus the line as something two-dimensional, and the body as a sum of surfaces and thus the surface as something three-dimensional, i.e. not as immaterial, and to treat it in the mathematical procedure with all success, although this interpretation was tantamount to sacrilege in relation to the Euclidean definition of these geometrical elements.[33]

3.2.4 It Is Not a Matter of Unscrupulousness

The change in the function of theory, expressed in its practical orientation, thus undoubtedly made it possible to endure the discrepancy between the "ideal" reality of theory based on high abstraction and the reality that appeared, and to attempt compromises. However, the problem described by Lorini could not be remedied by the theorists of modern times simply being less scrupulous than those of antiquity. The cooperation that developed between theory and practice was based on the problem of practice—as Lorini's quotation also documents—because the artisanal procedure of approximation, as indispensable as it naturally remained, failed in certain situations despite the most ingenious exercise. Practitioners turned to theory precisely because they expected it to provide them with a qualitatively different way of solving problems. For the theoreticians, of course, it was just as out of the question to be content with any kind of half-science. Geometry, with its axiomatic structure and the rigor of its deductive reasoning, remained for them the guideline against which theoretical work was to be measured. The fact that Euclid's *Elements*, which had been known to the Occident since the early twelfth century, was the most widely printed book in the world alongside the Bible until the beginning of the twentieth century, or that Newton apparently based his *Principia* on the systematic form of the *Elements*, thus indicating that he saw the completion of physical science in a systematization comparable to Euclidean geometry, provides an idea of the fascination that this hitherto unique science exerted on the West. Finally, for ideological reasons, it must have seemed unsatisfactory to the theoreticians of the rising

[33]Archimedes had also committed this sacrilege in his statics, which was based on the study of mechanical devices. But he knew that he would have lost his reputation in the scientific world of his time and was therefore careful not to publish this work, which is perhaps more important for the history of science than his mathematical works, beyond the circle of his trusted students (cf. Dijksterhuis 1956, 60). This only illustrates once more the distance between antiquity and a time when Leibniz, Newton and d'Alembert developed and used the infinitesimal calculus, which was not fully compatible with the basic assumptions of seventeenth and eighteenth century mathematics, because it made physical theories representable and applicable.—But even in the seventeenth century, this use of Euclidean geometry, which had been enforced since Benedetti and Galileo, remained subject to criticism. Galileo's student Torricelli, for example, defended himself against this criticism with remarkable clarity (cf. Dijksterhuis 1956, 400 ff.).

bourgeoisie to merely pragmatically blur the difference between theory and appearing reality, between concept and appearance. The dualistic worldview was always able to draw profit from this difference.

3.2.5 *The Connection to be Deciphered*

In no way, therefore, could it be a matter of covering up or ignoring this difference; progress could only be made if it was possible to discover the connection between the disparate; to understand concept and appearance in their difference as a necessary unity. The contact between theory and practice was once again of decisive importance in accomplishing this task, in which one can, in my opinion, see the core of the development of modern thought into science. In any case, the first clue to its solution was provided by the application of mathematics, namely geometry, i.e. the only science with a systematic form that had been developed up to that point.

3.2.6 *Application of Mathematics*

3.2.6.1 Astronomy

The early bourgeois period expanded the application of mathematics, as will be shown, but it by no means discovered this application. Mathematics had been established in practice from time immemorial, not only in its elementary form as an arithmetical operation with natural numbers, but also in its "heights", as axiomatic geometry; in the Middle Ages, the West had only to pick up the thread of antiquity again. At the center of this application of mathematics was astronomy, which even in its Ptolemaic form can be described as the science that was truly indispensable for material production at that time. Its very application in astronomy not only proved the usefulness of mathematics; it also seemed to open up a new dimension of practice, the dimension of systematic construction. No longer trying things out and ultimately trusting in chance like the craftsman's method, but proceeding systematically, deducing from a few axioms, taking the shortest and at the same time safest, necessary path, mathematics revealed a constructive potency which predestined it to become a model of revolutionizing practice. When, at the end of the fifteenth century, artists were able to use this constructive potency of mathematics in a central perspective in another field, mathematics acquired the status of such a model.

3.2.6.2 Central Perspective

In these cases of application, it was evidently possible to bridge the gap between the world of pure forms and appearing reality not only approximately, but completely. The application of geometry in Ptolemaic astronomy, which from the outset fundamentally contradicted the dualistic worldview, was to be tolerated by the latter because astronomy was dealing with the translunar sphere of appearing reality, i.e. the sphere for which the dualistic worldview assumed no difference between pure form and appearance.[34] From here, it is possible to gauge what it must have meant when the Renaissance, in central perspective, reconstructed appearing reality below the lunar sphere in its spatial-optical aspect with the help of geometric forms; the scientific concept had been brought back to earth, had been grasped as a concept of appearing reality.[35] This not only negated the dualistic context of geometry; it also succeeded in applying geometry not only in the singular and therefore seemingly exceptional case of astronomy. Nothing seemed to prevent it from being considered universally applicable.

The most important achievement, however, was that a giant step had been taken toward deciphering the constructive potency that was so fascinating about geometry and, within it, about science. In Euclidean geometry, deduction made higher-ranking concepts and forms give rise to others of lower rank; deduction remained a procedure limited to the "ideal" world of concepts and pure forms; objective spirit remained with itself. In the application of geometry to the reconstruction of the ordinary appearing world, the deduction from a system of general concepts admired in geometry revealed itself as a procedure for mentally reproducing the appearing world. The mind was able to allow appearing reality to emerge from concepts just as strictly and methodically as Euclid had done with regard to imaginary geometric forms.

In the process, the concept also took on a different character. It no longer represented a form of being that belonged to another world in which the concept contained it, while only traces and distortions of it were found in the appearing world. The concept now contained the general conditions and regularities from which appearing reality is to be reconstructed, from which appearing reality is to be brought forth. The connection between concept and appearance is thus decoded in a first form: the concept contains the conditions of generation of appearance; concept and appearance remain different things, but they form a necessary unity.

[34] In this context it becomes clear what a staircase joke it is in the history of science that Ptolemaic astronomy, with the assumption of eccentricity, only concealed the fact that it was also confronted by this translunar sphere of appearing reality with the problem of only being able to derive the appearances from the pure forms if it had previously developed the pure forms from the analysis of the appearances.

[35] Olschki referred to this aspect in his account of Brunelleschi (Olschki 1919–1927, vol. 1, 42 ff.).

3.2.6.3 Mathematical Speculations

In contrast to Euclid, deduction now no longer mediated two spiritual forms; but what this application of geometry shared with Euclid is that the general concepts as the basis of deduction were conceived of as spiritual entities. The appearing world seems to proceed from a spiritual one; it appears as an emanation of spiritual entities. Thus the dualistic form of objective idealism is overcome, but not this idealism itself, which has only assumed a pantheistic form. The seemingly fantastic speculations, especially of the Italian Renaissance towards the end of the fifteenth century, which attempted to uncover the principle of appearing reality in a systemic harmony of the proportions of forms (golden section), and to grasp this reality as a universe, make this just as clear as the philosophy of Nicholas of Cusa, who probably not coincidentally symbolized the idea of a unity of finite (conditional) and infinite (unconditional) reality in ingenious geometric speculations.

3.2.6.4 The Mysterious Origin of Mathematical Forms

The second section pointed to the historical-social causes of the half-baked way in which the feudalistic worldview was overcome in the early period of bourgeois development.[36] As far as the immanent theoretical development is concerned, it must first be taken into account that this important breakthrough in the application of scientific theory involved the application of a system of concepts that the Renaissance had not developed itself but had found in the legacy of antiquity.[37] This means, however, that the origin of the abstract concepts, from whose system it was possible to deduce so excellently, lay in the dark.

Insight into their genesis was already difficult due to the historical discontinuity between antiquity and the West. But even if the theoretical achievements of antiquity had not been recovered in a roundabout and adventurous way as treasures of an exotic culture that had sunk like Atlantis, in fragments and predominantly only in the form of their results, the insight into the genesis of geometry and its concepts would have faced scarcely less difficulty. For the Greeks, like their occidental contemporaries, marveled at the geometry that they had brought to the height of a scientific system as a miracle. In the writings of the Greeks, especially in those on history and geography, there are indeed references valuable for the history of science to the Greek inheritance of the Babylonians and Egyptians and, in connection with this, to the knowledge that geometry arose from the art of field measurement. But their transition from a practical art of measurement to a science that made its

[36] Regarding the Italian Renaissance, cf. Tomberg (1973, 57–76).

[37] The fact that the application of Euclidean geometry in the construction of the central perspective also required the development of some new, above all non-geometric, procedures is certainly of extraordinary importance with regard to the further development of mathematics. However, in my opinion, this does not change the fact that it was essentially a matter of applying a found conceptual system.

deductions on the basis of a system of concepts did not appear to them as a continuation or a leap, but as a break with this origin of geometry. At the scientific level they had reached, they regarded geometry as the science of spiritual entities founded in themselves. The social causes of this conception of geometry have already been mentioned. A class that looked down contemptuously on work, whose distance from the confrontation with the reality that appeared in practice was a constitutive part of its self-understanding and thus of its worldview, did not have the prerequisites to understand the result of geometric abstraction precisely as the result of a complicated process of abstraction, as a side of reality worked out by this process. How little the process of abstraction was understood can best be studied in Plato, who famously portrayed it as a process of anamnesis, in which the opposite of anamnesis, namely the repression of the real process in which abstract concepts such as the definitions and axioms of geometry arose, is eloquently expressed.[38]

3.2.6.5 Evidence

Euclidean geometry admittedly favored the misjudgment of the process of abstraction because its definitions and axioms, once found, actually seem evident.[39] An idealistic understanding of the general concepts forced itself upon us, so to speak. If the apparent evidence of general concepts could generally not yet be understood in a subjective idealist way under the social conditions of the early bourgeois epoch, it had to reinforce the objectively idealistic view that general concepts reflect spiritual entities.[40] The standpoint, achieved above all in the construction of the central perspective, of grasping scientific derivations such as those of geometry as reconstructions of appearing reality, and thus the general concepts as reflections of the general conditions and lawful forms from which appearing reality emerges, thus had to remain a half-achievement. It is true that in general no one yet imagined that sciences in the manner of geometry possess their unique systematics and constructive potency because the human mind creates appearing reality according to its principles. But the scientific derivation appeared to trace a magical form of production insofar as it seemed to reconstruct the production, the emanation, of appearing reality from spiritual entities.

[38] The mechanical-materialistic way in which Aristotle imagined the formation of general concepts (cf. above all *Posterior Analytics* II 19) could hardly form a counterweight, because it also contains traits of an intuitive epistemology.

[39] Augustine provides a good example of how momentous the apparent evidence of mathematical abstractions at this early stage of mathematical development was for epistemology. He based his theory of illumination, which played a considerable role beyond the philosophical debates of the thirteenth century, on the evidence of mathematical forms, among other things.

[40] Exceptional examples such as Nikolaus von Autrecourt show, however, that this form of processing was already in preparation.

3.2.6.6 Magical Production

This appearance had to arise—apart from the social conditions, as I said—not only because the origin of general concepts such as geometry lay in the dark, and their apparent evidence pointed to something like participation in spiritual entities. The fact that a systematic reconstruction of some aspects of appearing reality from a system of abstract concepts in astronomy and central perspective succeeded because these concepts reflected moments of material reality in a general form was hardly to be unraveled, above all because the reconstruction in these cases of application of theory concerned only the spatial relations of appearing reality, i.e. only a dependent condition of form of its reality. It was not appearing reality that was reconstructed in these cases, but only a formal aspect of appearing reality; it was therefore not a matter of a mental reproduction of a real process of generation from general conditions. To reconstruct appearing reality from such of its general moments from which it cannot really emerge had to fortify the appearance that the systematic and constructive potency of scientific deductions is based on the reproduction of magical processes of generation.

3.2.6.7 Application in Technology

This appearance had to dissolve in the attempt to use the constructive potency of science for ordinary, earthly production, which is only able to produce something from material conditions. The difficulties which inevitably came to light in this process had to lead to the elimination of this appearance with the discovery of the real foundations of the constructive potency of science. These difficulties already became apparent when the practitioners, together with the theoreticians, set about expanding the application of mathematics.

The application of mathematics initially increased with the expansion of the trade in goods. However, accounting requirements and the need to offset the diversity of currencies and measures, which is almost inconceivable today, not only led to a quantitative expansion of arithmetic practice; these requirements also gave rise to arithmetic problems for which practicable and easily manageable solution techniques had to be developed first—for example, for dealing with decimal fractions, without which the offsetting of currencies and measures in the general ledger would only have caused the conspicuous variegation that characterized the political map of Europe at that time. From the fifteenth century onwards, however, mathematics was also increasingly applied to solving technical problems, initially above all in war technology revolutionized by gunpowder (fortress construction and ballistics), but then also in civilian production (shipbuilding, harbor and canal construction, lifting machines, drive and first working machines). Here, too, it was not only a matter of a quantitative expansion of mathematical practice, but mathematics was forced to develop further if it wanted to meet the tasks posed by practice; in particular, the problems posed by ballistics demanded the development of a mathematical field that, in the form adopted from the Arabs, could still almost be described as new mathematical territory, algebra. But that was not all. In this application of mathematics to the solution of technical problems, difficulties arose whose resolution could not be expected from further inner-mathematical progress alone; difficulties in which the applicability of mathematics to the solution of technical problems made the scientific foundations of technology a question.

An engineer who had to build fortresses corresponding to the progress of firearms inevitably experienced the limits of the blessings of mathematics.[41] For example, once the engineer knew the angle of inclination of the wall, which represented the best compromise between its two tasks—to be impregnable and to withstand the projectiles by softening their impact—mathematics was of great use to him for construction and building. The necessary translation of this angle, for instance, into the relation between the intended height of the wall and the horizontal distance by which the wall begins on the ground plane before the height perpendicularly dropped from the top of the wall, could, of course, also be achieved by trial and error; but at what expense compared with a geometrical derivation. However, for the question of the extent to which different angles of inclination defuse the effect of the projectile, mathematics could at first at best be used by measuring procedures for the unambiguous linguistic fixation of empirical experience, enabling exact communication in cooperation. However, the defusing effect of different angles of inclination itself could not be derived mathematically, unless it was assumed unchecked that it would increase steadily with the deviation from the perpendicular.

3.2.7 The Crucial Problem: Analysis

The problem at stake here is familiar. The constructive potency that geometry demonstrated in astronomy and in the construction of central perspective was based on the fact that Euclidean geometry had studied spatial relations at the level of abstraction of their pure form, discovered their regularities and represented them systematically.[42] Where spatial relations were concerned, as in perspective or Ptolemaic astronomy, conclusions could be drawn from Euclidean laws with the most certain success. But as soon as it was a question of relations other than space, axiomatic geometry could not be used deductively, or only in those cases in which these other relations, for whatever reason, were structurally identical with those of space. To decide this, however, presupposes the analysis of these non-spatial relations, and indeed an analysis that has reached the level of abstraction corresponding to that of pure spatial forms in geometry. The problem of analysis becomes central.

With this, theory reached the stage of its breakthrough to scientific form. When the methodological significance of analysis became apparent, the constructive

[41] Cf. Hall (1964, 370).

[42] The relativity of the concept of space underlying Euclidean geometry, as revealed by Riemann and Lobachevsky, is of course irrelevant to our problem.

potency of science could finally be deciphered to such an extent that, regardless of the difficulties that still had to be overcome, the way in which the promise made in this constructive potency of practice could be redeemed became visible in principle. Deduction could then be understood as the mental reproduction of a production process comparable to ordinary material production, because it was stripped of any semblance of magical production by the analysis that preceded it. The analysis had to find in appearing reality the moments from which something real actually emerged, from which it was to be produced. Deduction no longer had anything mysterious about it; appearing reality was to be deduced, to be brought forth, from a system of general concepts, because these general concepts reflected what analysis had discovered in appearing reality as the general conditions of production of that reality.

As long as this function of analysis had not become fully conscious, the general terms remained ambiguous. After the dualistic understanding of general concepts had been broken through the application of mathematics in astronomy and in the construction of central perspective, appearing reality was the criterion of truth of these concepts insofar as it could be derived from them. However, the evidence or derivability of these concepts from evident concepts was regarded as a sign of their truth. Without insight into the meaning of analysis, the formation of concepts could therefore not be consistently distinguished from arbitrariness and speculation. As before, an alternative to each concept could be constructed, which could claim that appearing reality could also conceivably emerge from what it contained. Under these circumstances, the criterion of decision could not lie with appearing reality, but only in something like the rules of the theoretical procedure. The demand for evidence or the characteristic "claire et distincte," which Descartes formulates as a proof of the truth of general concepts from which deduction can be made, are such ultimately speculative rules as the already mentioned one that preference is to be given to the concept or system of concepts that allows a phenomenon to be deduced more simply, more "economically". The insight into the methodological character of analysis did not make these prescriptions and principles all completely worthless, but it did destroy their claim to be able to decide on the truth or falsity of a specific conceptualization. Analysis made it the criterion of truth that the conceptual system from which deduction can be made has—as Newton puts it—a "manifest" character, i.e. that it is derived from the analysis of appearing reality and reflects the moments of appearing reality from which a real could not only possibly be derived, but can actually and demonstrably be derived, i.e. really brought forth.

3.2.8 Newton and the 'Occult Qualities'

"These Principles I consider (here: mechanical forces)," Newton writes,

> not as occult Qualities, supposed to result from the specifick Forms of Things, but as general Laws of Nature, by which the Things themselves are form'd; their Truth appearing to us by Phaenomena, though their Causes be not yet discover'd. For these are manifest

> Qualities, and their Causes only are occult. And the Aristotelians gave the Name of occult Qualities only as they supposed to lie hid in Bodies, and to be the unknown Causes of manifest Effects: Such as would be the Causes of Gravity, and of magnetick and electrick Attractions, and of Fermentations, if we should suppose that these Forces or Actions arose from Qualities unknown to us, and uncapable of being discovered and made manifest. Such occult Qualities put a stop to the Improvement of natural Philosophy, and therefore of late Years have been rejected. To tell us that every Species of Things is endow'd with an occult specifick Quality by which it acts and produces manifest Effects, is to tell us nothing: But to derive two or three general Principles of Motion from Phaenomena, and afterwards to tell us how the Properties and Actions of all corporeal Things follow from those manifest Principles, would be a very great step in Philosophy, though the Causes of those Principles were not yet discover'd: And therefore I scruple not to propose the Principles of Motion above-mention'd, they being of very general Extent, and leave their Causes to be found out.[43]

Newton thus criticizes precisely this conceptualization, which is not immune to arbitrariness and speculation, and at the same time clarifies the decisive point in which the conceptualization of real science differs from it: "But to derive two or three general Principles of Motion from Phaenomena, and afterwards to tell us how the Properties and Actions of all corporeal Things follow from those manifest Principles…". A little later he calls the method by its name: "As in Mathematicks, so in Natural Philosophy, the Investigation of difficult Things by the Method of Analysis, ought ever to precede the Method of Composition. This Analysis consists in making Experiments and Observations, and in drawing general Conclusions from them by Induction, and admitting of no Objections against the Conclusions, but such as are taken from Experiments, or other certain Truths. "[44] To this can only be added that the experiment is not only analytical in character, but in its function of proof it is precisely deduction that became material production.

3.2.9 *Elaboration of the Analytical-Synthetic Method*

The methodological elaboration of the analytic-synthetic (inductive-deductive) character of natural science can be traced back to the fourteenth century, whereby the heterodox Averroists and the Alexandrians of Padua and Bologna, whom we have come to know as pioneers in the cooperation between theoreticians and practitioners, were also the pioneers of this methodological clarification.[45] In this

---

[43] Newton, *Opticks*, Book III, Chap. 1, Query 31, in Newton (1952, 401).

[44] Ibid., 404.

[45] See note 9 of this chapter. Zabarella's *De natura logicae* and *De methodis* are only the preliminary final summary of this methodological explanation of the Paduans, which can be traced back to Pietro d'Abano.—At this point, I think it becomes strikingly clear how confusing the widespread view is that modern science had to overcome Aristotelianism par excellence. The analytic-synthetic method can be seen as preformed in the concepts ἐπαγωγή-ἀπόδειξις and their elaboration (on the differences in detail cf. e.g. Kullmann 1965). The contribution to the genesis of modern science made by heterodox Averroism, which—to put it simply—is a further development of the materialistic side of Aristotelian philosophy and at the same time a demarcation and criticism of its

clarification of method, which is still inadequate in some respects compared to the Galilean version of the "metodo resolutivo" and the "metodo compositivo", it is of particular interest that the Paduans—in contrast to Aristotle—rejected the requirement of evidence for the terms and propositions determined by risolutio, from which conclusions are to be drawn compositively, and thus an axiomatic natural science in the Euclidean sense.[46] Precisely this detail seems to show that the Paduans developed their doctrine of methods on the basis of insight, on which the speculative character not only of the feudalistic, but likewise of the undeveloped bourgeois theory was primarily based. The shortcomings of their methodology are primarily due to the fact that the method basically remained a postulate, no matter how much the Paduans strove to apply their method in their scientific research, especially in medical research; the reason for this, which will be discussed in more detail in a moment, lies in the fact that the level of abstraction of the concepts to be gained resolutively does not depend solely on intellectual conditions. How precisely the Paduans had hit the decisive point for the further development of the theory with their methodology, however, was confirmed in the seventeenth century. The scholars of the seventeenth century never tired of emphasizing the analytical character of the new science.

It is obvious that with the elaboration and conscious application of this method—which had of course long been practised semi-consciously and without systematics—something was gained for the scientific procedure, and not only technically. Nor is the significance of this methodological clarification limited to the fact that with it the theory had in principle acquired the form of science. What constituted its scientific form also had extraordinary significance for the ideological struggle of the rising bourgeoisie. According to this method, the truth of a theory was based on the fact that in it appearing reality was explained by its derivation from such "principles" (Newton), which for their part were derived in the analysis from appearing reality, which thus represented the mental reproduction of certain moments of appearing reality. Secondly, by proving the truth of a theory by the fact that the explanatory derivation of appearing reality from these "principles" was ultimately subjected to the test of proving the theoretical derivation of a thing by its practical production from the moments of appearing reality reproduced in thought in the "principles", this theory redeemed the assertion directed against the feudalistic world view, namely that appearing reality represents a connection of inner necessity

idealistic side, can hardly be overestimated and shows that modern science is also based on materialistic Aristotle. Thus it cannot be regarded as a coincidence that R. Grosseteste and R. Bacon, who decisively promoted the reception of Aristotle in thirteenth century England, were at the same time pioneers of scientific methodology by distinguishing between inductive, experimental and mathematical knowledge and attempting to present them in their context. This does not deny that modern natural science, especially physics, had to overcome Aristotelian natural philosophy in terms of content. But what it had to overcome first and foremost was the fusion of the idealistic side of Aristotelian philosophy into dualistic Platonism, which in my opinion had been the core of scholastic philosophy since Thomas.

[46] Cf. Heimsoeth (1912, Part 1, 17 pass.), esp. 20 f.; cf. especially the first book of the *Posterior Analytics*.

founded in itself, to be explained from itself. In the scientific form of this theory, the bourgeois world view simultaneously stripped itself of the speculative form that contradicted it.

3.2.10 Preconditions for Its Practice

With the elaboration of the analytical-synthetic method, however, the way was only paved in principle for a theory of nature with a scientific character and a form of constructivity appropriate to material production. Establishing a method and being able to realize a method in the scientific process are two different things. It is no coincidence that the step from the first elaboration of the analytical-synthetic method to its successful application in physics took about 200 years. Realizing the analytical program set up by this method represented one of the greatest difficulties to be mastered.

"To derive two or three general principles of motion from phenomena," as Newton's formulation quoted above reads, means very different things depending on what kind of phenomena we are dealing with. It would be possible, in my opinion, to divide the sciences according to the difficulty which each class of phenomena presents for the analytical discovery of the "principles" governing them, and to see in this an explanation for the chronological order in which the individual sciences reached the stage of a systematic science. It is undoubtedly easier to abstract the forms of space from the phenomena than the forms of mechanical force. Both seem—*post festum*, of course—almost child's play compared with the discovery of the "principles" of chemistry, physiology, biology, etc., for these sciences have to observe and compile an enormous abundance of phenomena, sift through them and try to classify them in ever new attempts, before it begins to emerge at which points analysis could identify connections from which the "principles" of the field are to be inferred. Compared to these disciplines, it can be said that geometry and mechanical physics were presented with the phenomena whose analysis could lead to the discovery of the "principles" on a platter, as it were, by material practice: measuring instruments such as measuring line, square and compass, or mechanical devices such as lever, inclined plane, pulley, centrifuge etc. Moreover, it is only a few phenomena whose analysis leads these disciplines to the discovery of their "principles". It is therefore no wonder that the physicians of Padua, for example, who played a decisive role in clarifying the analytic-synthetic method, were unable to make any progress in their field, nor that chemistry initially did not go beyond speculative attempts at systematization. But it is also no wonder that mechanical physics was only able to solve its comparatively "simple" task with the greatest effort; after all, since it could learn nothing or rather only misleading things from the analytical process underlying geometry for the reasons mentioned above, it had to carry out pioneering work for all scientific disciplines, had once again to find the way in which "two or three general principles of motion" could be derived in a scientific manner from the phenomena. One of the main tasks to be accomplished here was the development of a critical empiricism suitable for this purpose.

This critical empiricism could only be developed, as discussed above, by pushing beyond the "shallow empiricism" of the practitioners. This "shallow empiricism" was based on the one hand on the casuistic disposition of the practitioners at this stage of the development of production, and on the other hand on the fact that the state of the productive forces imposed limits on the practitioners' abstraction of certain phenomena (e.g. friction); abstraction presupposes the technical management of such phenomena. The technical management of a phenomenon such as friction, on the other hand, presupposes that it is investigated separately, presupposes a structuring, abstract empiricism. This, of course, is by no means alien to the practitioners within the limits set by the state of development of the productive forces. On the contrary, it must be said that the critical empiricism of modern natural science, however much it had to be developed at a distance from the practitioners, was at the same time built on that of the practitioners. Material production, as shown in the first section, created the preconditions for analytical-synthetic science by providing it with the necessary instruments and materials, and by creating inexhaustible stimuli for scientific research with its problems; above all, however, it prepared the use of natural potencies in a general form by working out, in the extension of the materials used (substitutive) and in their formation (isolative), the possibilities of distinguishing between the natural things used and the effect for the sake of which they were applied. In this way, however, it directly prepared the way for scientific analysis and abstraction.[47] In the exceptional sectors of production in which machines were already being used, in which the formation was no longer determined by the capacity of the bodily organs of human labor, as it was in manual production, these effects could finally be studied unimpaired by the special 'natural thing' of human *physis* or body, and the possibility arose of acting not through "the powers of the hand, but through those of nature" (Newton).[48] Without these preconditions of abstraction created by material production, analytic-synthetic natural science would have had to remain an intention, and precisely also the critical empiricism demanded of it. This is also one of the reasons why a scientist like Galileo, who was fully aware of the contradiction between his theoretical work and the more immediate cognitive needs of practitioners, was not for a moment in danger of falling back into speculative theory, but saw in theoreticians of his kind and in practitioners, despite this contradiction, two necessary sides of the one party opposed to speculative theory; in the three characters of the *Dialogo* and the *Discorsi* Galileo gave unmistakable literary expression to this conviction.

---

[47] Cf. 24 f.

[48] Preface to the *Principia* (trans. modified), in Newton, 1779–1785, vol. II, x; Newton (1999, 28).

3.2.11 A Step Beyond Practice

The contradiction that arose between theoreticians and practitioners with regard to empiricism did not result from two fundamentally incompatible attitudes to empiricism. Rather, it arose because theory, in order to find its "principles", had to reach beyond the level of abstraction that was attainable to practitioners due to the state of development of the productive forces. The principal limit that the state of the productive forces at that time imposed on practitioners in terms of abstraction and conceptualization consisted, as the first section showed, in the fact that in material production, the formative forces of nature could not yet be utilized in a general form, i.e. as such, even though developments were taking place in material production that would later prove to be the preparatory steps of such utilization. Material production was thus not able to produce things from its general conditions; it still only mastered formative potencies of nature in the form of effective properties of certain natural things. But it was precisely this limit that theory had to cross in order to become a systematic science; its "principles", in order to be truly "principles", could no longer reflect the forces of nature as properties and modes of action of some or many natural things, but as natural forces themselves, i.e. in general form. Material production, as we have seen, offered theory all the conditions for taking this decisive step; but it could not, at its then stage of development, accompany theory on this path.

3.2.12 Property or Moment?

For the purposes of this work, a detailed discussion of this step in the philosophy of science is dispensable; it is sufficient here to consider, as briefly as possible, what seems to me to be a fundamental aspect of the problem posed to theory in the attempt to find the "principles" of the modes of action of natural things: the forces of nature that are effective in them. This aspect can perhaps be characterized in principle form in the following way: natural forces act only as effective properties of natural things. The natural thing and its conditions contain the reason that certain effective properties are found in the natural thing in question; but the mode of action of these effective properties cannot be established on the basis of the natural thing and its conditions. The mode of action of these properties can only be deciphered from the interrelationship that they form with each other and/or with the properties of other natural things. This causal relationship contains the reason for the mode of action of the effective properties of natural things involved in it. The general form of this causal relationship must be identified in order to find the natural force that expresses itself in the modes of action of the properties. This requires that the interacting properties of natural things be understood not as properties of things but as moments of their causal relationship, i.e. as things that condition the causal relationship and are conditioned by it.

The separation of the properties of natural things, which are, or insofar as they are, moments of a particular interaction under investigation, is only one difficult step that must be taken. Their conception as moments instead of as properties—, the preconditions of which were dealt with in the first section and in which the specificity of the generalization required here lies—is more difficult still. As a property or, in this case, state of a natural thing, it makes a difference whether it moves with respect to a point on the surface of the earth or is at rest. As a moment of the interaction into which bodies enter with one another due to inertia, both states are the same (the third definition and the first axiom in Newton's *Principia*): what is its weight as a property of a body (it "has" weight) is its mass as a moment of the relationship of attraction between bodies (they have an attractive effect as masses), etc. The effort that this distinction, this generalization, required is well known, and everyone whose profession is not physics knows the trouble caused by the mere comprehension of the solution found.

This aspect is significant in the context of this work because it seems to me to expose the core of the problem that has so far been shown as the opposition between critical and shallow empiricism, and also of the so-called idealization theorem. With this step towards generalization, theory seems to detach itself from appearance and to investigate a fictitious world, because it does not take the aspects of appearance under consideration as what they appear as, namely as properties of things, but as moments of a connection that does not appear as such, but is inferred mentally from the appearances. Of course, this has nothing to do with idealization, for the forces of nature do not by any means only work in an ideal world in the regularity investigated by theory. Apparent deviations from this regularity with which they act in appearing reality are due to the fact that no natural force acts in appearing reality in isolation, without the intermediary of other natural forces modifying its effect.[49] This empirical theory is critical because it does not take appearance for reality, but distinguishes in appearing reality between appearances and the forces at work in it. But it is even more important that it does not stop at this distinction, which would leave appearances as they appear, but sees the appearances differently on the basis of the knowledge of natural-law relationships, namely as moments and results of the recognized effective relationships of nature. That this is not simply a different view of appearances, but their true conception as is demonstrated by this theory experimentally, whereby it allows certain properties of things to act as moments of the process from which other appearances emerge as results.

[49]The extent to which the effect of these other forces of nature can be determined in individual cases depends on the stage of development of material production; here lies the historical aspect of the 'idealisation' problem already alluded to.

3.2.13 *Quantification or Determination of Relationship?*

This critical, or better, *scientific* conception of appearance corresponded to a conceptualization—one presupposes the other—that was fundamentally different from the metaphysical (cf. the second section) conceptualization that had prevailed until then. The latter, no matter to what heights of abstraction it ascended, had never really left the level of appearance, as its predicative structure shows. The generality that its concepts represent is a merely nonsensual thing with merely nonsensual properties. The 'principles' that contain natural forces could not be formulated with such a conceptualization. Rather, concepts were required that express a general interrelation of effects, concepts that grasp the interacting as elements and their interaction as their relation, i.e. a form of conceptualization that does not predicate something of something, but reflects relations in their general form. The 'principles' that the theory had to find could only be expressed conceptually as relations. In my opinion, this is the meaning of Galileo's words, quoted everywhere, that the "book of nature" was "written in mathematical language" and could therefore only be deciphered if one was proficient in this language.[50]

3.2.14 *New Mathematical Speculation?*

Galileo is not advocating a renaissance of Pythagorean speculations, he does not at all mean that reality can be reduced to quantity or geometric forms. His entire work speaks against this.[51] But what is Galileo saying, if not that the laws of nature can only be understood as ratios and thus expressed in ratio determinations, ratio determinations as they are peculiar to mathematics, as they were known to him only from mathematics in non-speculative form.[52] Newton formulated it as follows: "… as all these things (laws of optics) follow from the properties of the light *by a mathematical way of reasoning* (W. L.) …";[53] and in the *Principia*: "Mathematics requires an investigation of those quantities of forces and their proportions that follow from any conditions that may be supposed. Then, coming down to physics, these proportions must be compared with the phenomena, so that it may be found out which conditions

[50] Galilei (1896) (vol. VI), 232 (*Il Saggiatore*, Engl. *The Assayer*).

[51] Cf. Strong (1936), which is a critical examination of Burtt (1925). This debate is reflected in Clark, op. cit., in Cassirer (1940), and in other authors of the 1930s and 1940s; This is still often overlooked in the widespread accounts of the history of science; cf. e.g. the contribution by R. Lenoble and Y. Belaval in Taton's *Histoire générale des sciences* (Lenoble and Belaval 1969, 201).

[52] With regard to Descartes, E. Cassirer comes to the same conclusion (Cassirer 2005, 11/1939, 20): "It [the *mathesis universalis*] is now no longer about any particular of any particular individual content, but is the general science of relations, in so far as they are capable of exact expression" (*Ed. note*: transl. added).

[53] Newton (1952, 240).

[or laws] of forces apply to each kind of attracting bodies";[54] one could even formulate this as follows: If one comes down to physics, one must analyze the phenomena in order to find out with which mathematical structure their lawful correlation of effects can be expressed.[55]

The reasons for this ahistorical and thus still metaphysical conception of nature, which is satisfied by mathematical determinations of relations, will be discussed in the next section. More interesting than the one-sidedness of classical natural science at this point is first of all that men like Galileo were sure not to fall back into speculation with this "mathematization" of nature. After all, this is by no means self-evident after the use of mathematics as it was common before Galileo and Kepler. When, for example, Bacon, who found the most concise formulations against the Peripatetics, in a quite Aristotelian fashion wants mathematics in natural science to be limited to a subordinate role, this by no means expresses ignorance, but his fear that otherwise the materialist character of natural science could be endangered; or, if it does express ignorance, it is at least the same "ignorance" that Huygens or Leibniz showed a generation later in their rejection of Newton's gravitation.[56] What set them against gravitation acting through empty spaces, i.e. without mediating material, was the fear that with gravitation Newton had rehabilitated the assumption of spiritual forces in the theory of nature, from which they thought they had been happily expelled.[57] Galileo and Newton's experimental verification of their statements gave them the certainty that their theory did not rise speculatively above

[54] Newton, *Principia*, Book I, Section XI, Scholium in Newton (1999, 235).

[55] These remarks only touch on the so-called quantification of modern natural science; a detailed discussion, however, falls outside the purposes of this work. As far as Newton's mathematical-empirical method is concerned, there is, as it seems to me, much perplexity in the common literature on the history of science; cf. e.g. Dugas and Costabel (1969, 282 ff.), Hall (1963, Chaps. 3 and 4). This becomes clear when one observes that even the relevant studies that deal explicitly with Newton's method give more information about the epistemological, methodological and scientific-theoretical position of the respective authors than about Newton's theoretical concept: cf. in particular the volume edited by Butts and Davis (1970) (Buchdahl's contribution should perhaps be excluded from the criticism made here); as well as Wagner (1969), of the older works cf. Cassirer (2003, Chap. 2), Cassirer (1951, 53ff.), Cassirer (1999, seventh book, 329 ff.), Koyré (1965).

[56] Cf. especially the well-known passage in *De Augmentatis Scientiarum*, II, in Bacon (2000, 87f.).

[57] If any proof at all was needed that Newton, despite his speculations about the activating and reactivating effect of the Christian God on nature (cf. e.g. Newton 1952, 403–404), then in my opinion it proves his theory of gravitation, which under the circumstances of the seventeenth century rightly appeared questionable, and which at that time could only be put forward by a scientist who was absolutely certain of his materialistic view of nature. With regard to these speculations, it must be remembered that the alternative for Newton, given the state of physics at that time, consisted in cosmogonic constructions such as those developed by Descartes in his *Principles*; Newton rejected these as incompatible with natural science. He saw the achieved scientific level of natural theory as less threatened by Christian doctrines of faith (English style!) than by speculative theories such as Cartesian. This can perhaps be regarded as further evidence that the separation of faith and knowledge, which was founded in the hostility to science of Christian philosophy and was not really overcome by Thomism, was a more favourable condition for the development of science than the confrontation of two theories with a claim to scientific truth would have been.

appearing reality, but represented scientific knowledge of this appearing reality.[58] However, they could just as well—or more precisely: in connection with experimental verification—have invoked the difference that distinguished their "mathematical" conceptualization from the conventional one.

3.2.15 The New Science

It is precisely the difference between the conceptualization of modern science and the metaphysical conceptualization of theory that had prevailed until then that makes it clear that the observed distance of modern science from practice is not a matter of the old elevation above appearing reality, but rather, as it were, of surpassing practice, of hurrying ahead to a stage of practice whose realization in material production still required the developments of more than a hundred years. The "principles" had to contain determinations of relations because they were not about the realization of "true being", of "ideas" and spiritual entities, but about general interrelationships of nature. To explain a thing theoretically—that is the epitome of modern science appearing in classical physics—means to know from which general conditions it is to be produced according to objective laws. Modern natural science thus represents the intellectual conditions of a higher stage of material production, and only in this respect does it temporarily distance itself from it. With the industrial revolution, this temporary discrepancy between theory and practice is removed again. But this could only happen because material practice in the process of the formation of modern natural science was not only conducive to it in some respects, but had shaped and transformed theory in its innermost structure, making it what science is in contrast to speculation.

References

Bacon, F. 2000. Oxford Francis Bacon. In *The advancement of learning*, ed. Michael Kiernan, vol. 4. Oxford: Oxford University Press.

Bacon, F. 2012. Oxford Francis Bacon. In *Early writings 1584-1596*, ed. Alan Stewart and Harriet Knight, vol. 1. Oxford: Clarendon Press.

Blumenberg, H. 1965. In *Das Fernrohr und die Ohnmacht der Wahrheit. Introduction to Galileo Galilei, Siderus Nuncius*, ed. H. Blumenberg, 3–72. Frankfurt am Main: Insel.

Blumenberg, H. 1985. *The legitimacy of the modern age*, trans. R. Wallace. Cambridge MA: MIT Press.

Burtt, E. A. 1925. *The metaphysics of Sir Isaac Newton: an essay on the metaphysical foundations of modern science*. New York: Columbia University Press.

[58] There is countless evidence for this; I refer here by way of example to Galilei, Discorsi, German edition, op. cit. 162, and Newton (1952, 240).

Butts, R. E., and J. W. Davis, eds. 1970. *The methodological heritage of Newton*. Toronto: University of Toronto Press.

Cassirer, E. 1940. Mathematische Mystik und moderne Naturwissenschaft. *Lychnos* 5 (1940): 248–265.

Cassirer, E. 1951. *The philosophy of the enlightenment*, trans. F. Koelln and J. Pettegrove. Princeton: Princeton University Press.

Cassirer, E. 1999. *Das Erkenntnisproblem in der Philosophie und Wissenschaft der neueren Zeit. Zweiter Band*, Gesammelte Werke. Vol. 3. Hamburg: Felix Meiner Verlag. (first ed. Berlin: B. Cassirer 1907).

Cassirer, E. 2003. *Philosophie der Aufklärung*, Gesammelte Werke. Vol. 15. Hamburg: Felix Meiner Verlag. (first ed. Tübingen: Mohr 1932).

Cassirer, E. 2005. *Descartes: Lehre–Persönlichkeit–Wirkung*, Gesammelte Werke. Vol. 20. Hamburg: Felix Meiner Verlag. (first ed. Stockholm: Bermann-Fischer 1939).

Crombie, A. C. 1952. *Augustine to Galileo: The history of science A. D. 400-1650*. London: Falcon Educational Book.

Da Vinci, L. 1955. *The notebooks of Leonardo da Vinci*, ed. & trans. E. McCurdy. New York: Braziller.

Dijksterhuis, E. J. 1956. *Die Mechanisierung des Weltbildes*. Berlin: Springer.

Dugas, R., and P. Costabel. 1969. Naissance d'une science nouvelle: la mécanique. In *Histoire générale des Sciences*, ed. R. Taton, vol. II, 252–288. Paris: PUF.

Feldhaus, F. M. 1930. *Die Technik der Antike und des Mittelalters*. Potsdam: Athenaion (Repr. 1971).

Feuerbach, L. 1969. *Geschichte der neueren Philosophie von Bacon bis Spinoza I*. Berlin: Akademie-Verlag.

Galilei, G. 1895. *Le opere*. Vol. 5. Firenze: Barbera.

Galilei, G. 1896. *Le opere*. Vol. 6. Firenze: Barbera.

Gilbert, N. W. 1963. Galileo and the School of Padua. *Journal of the History of Philosophy* 1 (2): 223–231.

Hall, A. R. 1963. *From Galileo to Newton. 1630–1720*. New York: Harper & Row.

Hall, A. R. 1964. Military technology. In *A history of technology, vol. 3. From the Renaissance to the Industrial Revolution, c. 1500-c. 1750*, ed. C. Singer et al., 347–376. Oxford: At the Clarendon Press.

Harig, G. 1958. Über die Entstehung der klassischen Naturwissenschaften in Europa. *Deutsche Zeitschrift für Philosophie* 6 (3): 419–450.

Harig, G. 1964. Galileo Galilei. *Einheit: Zeitschrift für Theorie und Praxis des wissenschaftlichen Sozialismus* 19(3).

Heimsoeth, H. 1912. *Die Methode der Erkenntnis bei Descartes und Leibniz*. Giessen: A. Töpelmann.

Horkheimer, M., and T. Adorno. 1947. *Dialektik der Aufklärung*. Amsterdam: Querido Verlag.

Horkheimer, M., and Adorno, T. 2002. *Dialectic of enlightenment*, ed. G. Schmid Noerr, trans. E. Jephcott. Stanford: Stanford University Press.

Hörz, H. 1974. *Marxistische Philosophie und Naturwissenschaften*. Cologne: Pahl-Rugenstein.

Husserl, E. 1992. *Die Krisis der Wissenschaften als Ausdruck der radikalen Lebenskrisis des europäischen Menschentums*. Hamburg: Meiner.

Kirk, G. S., J. E. Raven, and M. Schofield, eds. 1983. *The Presocratic Philosophers*. 2nd ed. Cambridge: Cambridge University Press.

Klaus, G., and G. Kröber. 1966. Mathematik und Philosophie in der Antike. In *Wissenschaft und Weltanschauung in der Antike: Von den Anfängen bis Aristoteles*, ed. G. Kröber, 69–121. VEB Deutscher Verlag der Wissenschaften.

Klemm, F. 1954. *Technik - Eine Geschichte ihrer Probleme*. Freiburg: Karl Alber.

Koyré, A. 1965. *Newtonian studies*. Cambridge MA: Harvard University Press.

Koyré, A. 1969. Sciences exactes. In *Histoire générale des Sciences*, ed. R. Taton, vol. II, 12–207. Paris: PUF.

Kristeller, P. O. 1961. *Renaissance thought. The classic, scholastic, and humanist strains*. New York: Harper Torchbooks.
Kristeller, P. O., and J. H. Randall. 1941. The study of the philosophies of the renaissance. *Journal of the History of Ideas* 2 (4): 449–496.
Krüger, L., ed. 1970. *Erkenntnisprobleme der Naturwissenschaften*. Köln: Kiepenheuer u. Witsch.
Kuksewicz, Z. 1965. *L'Averroïsme Bolonais au XIVe siècle*. Krakow: Zdzisław.
Kullmann, W. 1965. Zur wissenschaftlichen Methode des Aristoteles. In *Synusia. Festgabe für Wolfgang Schadewaldt*, ed. H. Flashar and K. Gaiser, 247–274. Pfullingen: Neske.
Lenoble, R., and Y. Belaval. 1969. La révolution scientifique du XVIIe siècle. In *Histoire générale des Sciences*, ed. R. Taton, vol. II, 195–216. Paris: PUF.
Ley, H. 1957. *Studie zur Geschichte des Materialismus im Mittelalter*. Berlin: Deutscher Verlag der Wissenschaften.
Ley, H. 1966. *Geschichte der Aufklärung und des Atheismus*. Berlin: Deutscher Verlag der Wissenschaften.
Maier, A. 1966. *Die Vorläufer Galileis im 14. Jahrhundert*. Rome: Edizioni Di Storia E Letteratura.
Newton, I. 1779–1785. In *Opera quae exstant omnia*, ed. S. Horsley, vol. 5 vols. London: J. Nichols.
Newton, I. 1952. *Opticks*, 4th edition (1730), ed. I. Bernard Cohen et al. New York: Dover.
Newton, I. 1999. *The principia: Mathematical principles of natural philosophy*. Translated by I. B. Cohen, A. Whitman, and J. Budenz. Berkeley: University of California Press.
Olschki, L. 1919–1927. *Geschichte der neusprachlichen wissenschaftlichen Literatur*. Heidelberg: Max Niemeyer Verlag.
Parmenides. 1922. Fragmente. In *Die Fragmente der Vorsokratiker*, ed. H. Diels, vol. 1, 4th ed., 148–165. Berlin: Weidmannsche.
Randall, J. H. (Jr.) 1941. Study of Rnaissance Philosophies. *Journal of the History of Ideas*, III/4, 449–496.
Schumacher, E. 1964. *Der Fall Galilei*. Berlin: Deutscher Verlag der Wissenschaften.
Strong, E. W. 1936. *Procedures and metaphysics - A study in the philosophy of mathematical-physical science in the 16th and 17th centuries*. Berkeley: University of California Press.
Tomberg, F. 1973. *Bürgerliche Wissenschaft*. Frankfurt am Main: Fischer Taschenbuch.
Wagner, F. 1969. *Neue Diskussionen über Newton's Wissenschaftsbegriff*. Munich: Verlag der Bayerischen Akademie der Wissenschaften.
Wolff, M. 1971. *Fallgesetz und Massebegriff: zwei wissenschaftliche Untersuchungen zur Kosmologie des Johannes Philoponus*. Berlin: De Gruyter.

Chapter 4
Contradictions Between the New Science and the Bourgeois Worldview as Reflected in Philosophical Treatments of Natural Science in the Seventeenth Century

4.1 First Section

4.1.1 Under the Spell of the New Physics

Classical physics, which emerged with Galileo and was worked out to a first conclusion and completion by Huygens and Newton, exerted an extraordinarily strong influence on the thinking of the seventeenth century. We must even ask whether there is another period in the history of ideas in which a scientific discipline left its imprint on all other sciences, on philosophy, and on general knowledge as the physics of the seventeenth century was able to do. This probably remains true even if we take into account that we usually think of the intellectual life of the seventeenth century as being determined by philosophers such as Bacon, Descartes, Hobbes, Spinoza, Leibniz and Locke, on whom I will essentially concentrate in this chapter, and thereby easily overlook the fact that the old way of thinking, which, despite all modifications, was characterized by scholasticism, was by no means overcome in the seventeenth century, but without doubt still dominated in most European countries.

In the Romance countries, intellectual development in the seventeenth century was under the shadow of the Restoration, not only in the form of the Spanish Inquisition, whose effects can be seen in the Galileo trial and Descartes' voluntary emigration. In Germany, the Thirty Years' War—not to mention the effects of the intellectual terror of a war officially fought over confessions—deprived cultural life of its ground simply by the collapse of all civil normality. Only in the last third of the century did German scholars in a few cities and at a few courts, who, under these circumstances, can hardly be regarded as characteristic of German intellectual life, recover the conditions to be able to participate in European development. Even in England, in the restoration period that followed the first revolution, it was not harmless to identify oneself without hesitation as a follower of, say, Hobbesian materialism. Finally, Spinoza, as much the Antichrist of the seventeenth century as the

W. Lefèvre, *Theory of Nature and Mode of Production*, Political Epistemology,
https://doi.org/10.1007/978-3-032-02295-0_4

accusation of Spinozism was tantamount to being denounced to the authorities, did not have to fear for his life in the Netherlands, perhaps the most free-minded country of the seventeenth century—except for the De Witt incident—but he did have to endure a life of isolation and suspicion. However, if scholars like Mersenne, Pascal or Arnauld, who were probably committed to counter-reformational Catholicism, were intensively engaged in the new physics and partly contributed essentially to its development, or if in Restoration England people like Bentley, Boyle or Hook undertook the almost bigoted attempt to support religion with the help of the new physics, then this underlines perhaps more than anything else the importance that classical physics had for the thinking of the seventeenth century, despite the intellectual-historical situation described.[1]

Despite the above-mentioned misgivings, then, we must agree with the literature on the history of science when it emphasizes the extraordinary effect of classical physics on the thinking of the time and states that classical physics was not only followed with attention but was often pursued directly by practically all the important scholars of the seventeenth century, since its specific content, mechanics, colored the idea—with A. Maier and Dijksterhuis, we can speak of a 'mechanization of the world view', whatever exactly must be understood by 'mechanization'—that above all the analytical/synthetic method proven in classical physics was of utmost importance for the understanding of theory in the seventeenth century. This fact itself need not be pursued further here; in the context of this investigation, we are rather interested in the importance that classical physics had for seventeenth century thought from a certain point of view.

4.1.2 Review and First Question

In the first chapter of this book we sought to convince ourselves that the upheavals of production, distribution and exchange accompanying the development of the bourgeois mode of production provided the conditions of development for classical physics, without which it could not have come into being. At the same time, however, we had to admit that these upheavals stimulated the development of modern natural science, but did not make it necessary or advance it because of compelling requirements of material production. As a motor in the genesis of modern natural science, so to speak, we tried in the second chapter to identify the ideological needs of the rising bourgeoisie, which it formed on the basis of the new social conditions in the struggle against the old social formation and its intellectual representatives. In doing so, we focused with regard to these ideological needs on the fact that it must be of decisive importance for the position, which conceives of appearing reality as a connection of inner necessity, whether this view is asserted merely speculatively

[1] Cotes' preface to the second edition of Newton's *Principia* (Newton 1713) can be regarded as an exemplary document for this.

or whether it is anchored in the realization of appearing reality itself.[2] For the overcoming of the speculative form of this ideological position, as it turned out, natural science played a key role before the eighteenth century; for at that time this position could not refer to the new social context arising with the development of the bourgeois mode of production, as much as it was its basis; if this context, due to the division of labor and the exchange of commodities, presented itself as governed by inherent laws, it was not one that existed autonomously as long as it had the intact feudalistic mode of production as a necessary condition of reproduction.

On the basis of these considerations, in this chapter we will first examine whether the interest that the philosophy of the seventeenth century showed in the achievements of classical physics can be understood as an expression of the attempt, with the help of the new natural science, to provide a non-speculative, scientific basis for the position, which could only be held speculatively as long as appearing reality must be understood as a connection of inner necessity, and thus to turn it against the feudalistic world view in a qualitatively new way.

[2] This formulation of the ideological position of the rising bourgeoisie, which was introduced in the second chapter, can easily cause misunderstandings in this chapter, which deals with the theories of certain thinkers; therefore some remarks on it. This formulation was chosen in the second chapter not least with a view to the reconstruction of the essential ideological contents from the conditions of real history; in doing so, however, I naturally had in mind the concrete articulations of bourgeois self-consciousness, the core of which the formulation attempts to grasp despite their diversity and contradictoriness.—Such an attempt is, of course, not without doubts: How should it be possible at all to establish a unified core of the natural-theoretical, economic, constitutional and legal-theoretical, religious and philosophical articulations; how should this be possible in view of the difficult problems of assigning the individual authors to the social groups, in view of the formation of camps from different strata in the social-historical movements, in view of the national peculiarities, etc.? How is this conceivable, when we generally have knowledge only of the thinking at the academies and high schools, but not of that of the 'common man', when even the spread and effect of 'high' thinking and its changes during such 'popularization' are little understood?—All these concerns were at the same time reasons for the reconstructive attempt of the second chapter. From the concrete articulations that have been handed down to us, I felt strengthened above all by the development of natural-theoretical thinking as well as by the continuity of a materialistic philosophy since the thirteenth century, which can be stated despite all contradictions, break-offs, retardations, "detours" and not least disguises. This formulation wants to express its content, and indeed in its specific contradictoriness, as this chapter will show. If the second chapter had tried to show in general and, in a few places, in exemplary form, how this formulation can prove itself as a thread of interpretation without doing violence to the complexity and contradictoriness of the concrete material, only this fourth chapter represents the 'test on the example'. In order to avoid misunderstandings, it must be emphasized that it is not a question of assigning individual theorists to this formulation, but—as I think—only a question of whether the philosophies considered here can be represented in such a way that they make essentials of their problems, their correspondence and their movement recognizable without any constraint. In this application, this formulation has to prove itself, whereby, by the way, it must not be forgotten that the treated philosophies of the seventeenth century are not examined here in general, but according to the question of this work in their relation to natural science.

4.1.3 Sober Welcome

First, it is important to note that the exuberant welcome of the new physics, an enthusiastic, *a fresco* realization of the possibilities it bestowed upon the bourgeois world view, as it happened to the Copernican astronomy by Bruno in the sixteenth century, is hardly to be found in the seventeenth century. Bacon, whose work undoubtedly shows features of such an emphatic welcome, is probably rightly assigned to the phase before the breakthrough of the new physics, despite the dates of his life, even if this classification by no means justifies him being dismissed—as often happens—as a dilettante enthusiast. Spinoza could probably best be called the Bruno of the seventeenth century because of the speculative character and mystical depth of his pantheistic-materialistic philosophy, even though the character of "consumptiveness" (Hegel) and asceticism inherent in his theory due to its orientation to geometrical deduction stands in the greatest conceivable contrast to the "a sensuous, poetic glamour" that characterized the materialism of the sixteenth and early seventeenth centuries.[3] If, however, a philosophy such as Spinoza's, which unlike Descartes' and Leibniz's is consistently and unreservedly speculative, is thought in the sparse or, to put it positively, disciplined terms of mathematical proof, then perhaps even in it a hint can be seen that the sobriety with which modern natural science was taken up in the seventeenth century expresses the transition of bourgeois thought from the speculative to the scientific form. That sobriety indexes this transition is clearly evident, for example, in Galileo. Versatile, interested, and active, knowledgeable and engaged in religious, philosophical, political, literary questions, permeated by the need to understand, operate, and propagate physics both through the central perspective, the confirmation of heliocentrism, as well as through the form of its literary mediation in the light of the ideological disputes of his time, Galileo proved to be a descendant of Campanella and Bruno. At the same time, however, Galileo's demonstrated methodological discipline, his self-imposed restrictions—most importantly, his conscious deferral of dynamic questions in favor of kinematic ones—and the demands he placed on scientific statements are of extraordinary importance for overcoming an emphatic, speculative materialism and advancing a sober, step-by-step, thoughtful, methodical way of thinking in the seventeenth century.[4]

[3] Marx and Engels, *The Holy Family*, MECW 4, 128.

[4] The materialism of the seventeenth century, which was oriented towards mechanical physics in its sparseness and one-sidedness is often regarded as a reduction and regression compared to, for example, Bruno, and it is supposed the contradictions of the bourgeois mode of production, which are expressed in the which were still concealed in the phase before the seventeenth century; cf. e.g. F. Tomberg (1973, 111) (We already dealt with such interpretations in Chap. 2, and we hope that the conclusion of this fourth chapter will contribute to the clarification of this problem). Regarding the narrowing of materialism in the seventeenth century, it is important to recognize in it the much more decisive event, that materialism, based on mechanical physics, begins to appear as a science. L. Feuerbach's interpretation still seems to me exemplary for the direction of an appropriate evaluation of the materialism of the seventeenth century, which is afflicted with all the shortcomings of mechanical physics; Feuerbach distinguished the criticism of 'mechanism' in the seventeenth century according to whether this criticism is developed "within and out of mechanism" or from a position that has not even reached the level of 'mechanism'; cf. Feuerbach (1848, 45).

The sober way in which classical physics was taken up and processed in the seventeenth century as compared with the sixteenth century thus hardly indicates that the new physics was not adopted and used in the sense of the ideological position of grasping and representing appearing reality as a self-grounded whole. On the contrary, it seems to confirm that this position begins to shed its speculative form and to develop a scientific shape on the basis provided by the new physics. In fact, there are abundant indications that the picture of nature emerging in mechanical physics was taken in the seventeenth century as a decisive confirmation of the ability to prove appearing reality as a self-grounded whole, and on the basis of this emerging picture of nature an attempt was made to work out a corresponding picture of the world. Such indications can be found above all in what must be understood at first by the 'mechanization of the worldview,' namely the attempts to explain the whole of appearing reality by means of the regularities investigated by mechanical physics and to understand it thus as a unity founded in itself.

4.1.4 'The Mechanization of the Worldview'

These attempts can be found, on the one hand, in the field of natural sciences itself, in the non-physical fields as well as in such fields of physics with respect to which it could by no means be taken for granted and unproblematic to work on them with the means of mechanical physics (e.g. optics, thermodynamics, magnetism). Harvey's discovery of blood circulation and the tendency, which followed this discovery, to regard human and animal organisms as complicated machines and to reconstruct their functioning according to mechanical laws and using geometrical derivation methods, belong here just as much as Boyle's attempt to give chemistry a scientific foundation with the corpuscular theory.

That in this respect one can rightly speak of a 'mechanization' of the worldview and not only of the view of nature is, of course, only shown by the attempts to explain also areas of appearing reality which do not belong to nature through mechanical physics and thus to give these areas a secure foundation for their scientific treatment. Hobbes and Locke must be mentioned here above all, who tried to secure a scientific starting point for psychology, social and state theory with a mechanical theory of drives and, in addition, started to base even epistemology on a theory of perception and abstraction, for which mechanical physics had constitutive importance.

Finally, the tendency to understand and to represent appearing reality with the means provided by classical physics as a self-grounded whole, as a unity scientifically explicable on the basis of its own laws, is probably most clearly expressed, despite all contradictions still to be discussed, in the corpuscular theories developed in the seventeenth century, which acquired a dominant influence on seventeenth century natural philosophy after Gassendi's Christian interpretation of ancient atomism had deprived them of the character of sheer atheistic provocation.[5]

How widespread this evaluation of classical physics was in the seventeenth century is shown by the fact that it is not only found among materialist theorists such as Hobbes, but also among philosophers such as Descartes and Leibniz, who cannot (Leibniz), or can only subject to the admission of contradictory elements (Descartes), be assigned to the materialist direction. The fact that Descartes and Leibniz were involved in the attempts to work out a cosmology on the basis of mechanical physics is not of decisive importance.[6] It is more decisive in this context that Descartes as well as Leibniz, although they considered a metaphysical foundation of appearing reality necessary—moreover, one that included God—nevertheless thought of appearing reality as structured by immanent, scientifically fathomable regularities. Thus, to put it more pointedly, God seems to be bound by the regularities inherent to appearing reality. This implicit materialist blasphemy was probably one of the main annoyances that the Cartesian "res extensa" aroused. In Leibniz, this idea emerges in his metaphysics itself, in that Leibniz elevates the harmony of the monads in the "monas monadum" to an independent existence, and at the same time seeks to ground this harmony in the monads themselves—in the mode of possibility.[7]

4.1.5 *Different Directions*

So there can be no doubt that classical physics in the seventeenth century was taken as a confirmation of the ideological position of the rising bourgeoisie and was used for the expansion of a materialistic world view of scientific character. But it is no less certain that this processing of classical physics did not happen uniformly, did not lead to any result generally accepted among the theorists, which can be understood from the position, that appearing reality must be understood as a context of inner necessity.

---

[5] Cf. especially the second volume of Laßwitz (1890).

[6] While Descartes worked out such a cosmology first in *Le Monde*, a work aborted under the impression of the condemnation of Galileo, and later such a cosmology is elaborated in his *Principia philosophiae*, Leibniz only has scattered approaches to it; from the writings are above all the '*Theoria motus abstracti*' and its counterpart '*Hypothesis physica nova*' (both in 1670), the '*Brevis demonstratio erroris*' (1686) and '*De ipsa natura sive de vi insita*' (1698). Since it is here only about the statement of the Leibniz's active interest in the elaboration of a cosmology on the basis of mechanical physics, we can disregard that the scattered and rather as moments of such a Leibnizian cosmology do not fit together seamlessly, but reflect the development of his thinking.

[7] Cf. Feuerbach (1848). Cf. König (1946)—It would certainly be interesting to investigate in another context how Leibniz's reproach (directed against Descartes' use of the ontological proof of God) that Descartes had neglected to prove first the possibility of God (cf. e.g. *Meditationes de cognitione, veritate et ideis* in Leibniz 1875–1890, vol. 4, 424), must be interpreted against this background

For a vivid visualization of this lack of uniformity, we need only remember that Descartes, who as author of the *Regulae*, of *Le monde*, of the cosmological parts of the *Principles* as well as natural scientist and mathematician certainly has to be counted among these theorists, presented his contemporaries by the dualistic confrontation of "res cogitans" and "res extensa" with a problem, the solution of which, despite the misgivings mentioned at the beginning of this chapter, could almost be called the dominating topic of philosophy in the second half of the seventeenth century.

4.1.6 Second Question

The overcoming of this dualism was a general program, whether of Hobbes, Spinoza, Leibniz or Locke. On the other hand, the question in which direction Cartesian dualism was to be overcome divided the philosophers, with Hobbes and Locke striving for a materialist, Leibniz for a metaphysical transcendence of dualism, while Spinoza stood in the middle insofar as both can be said of him. The fact that it was a general program to overcome Cartesian dualism could certainly be regarded to some extent as a further indication that the motive to substantiate and scientifically expand the conception of appearing reality as a coherence of inner necessity on the basis created by the new physics. However, Cartesian dualism itself as well as the contrasts between the attempts to overcome it point us above all to the fact that the initial question of this chapter does not lead far enough, that it is not sufficient to state that the new physics was processed in terms of the ideological position of the rising bourgeoisie in the seventeenth century. The contradictions and contrasts in this processing show, in any case, that the hope, which was significant for the genesis of modern natural science, to elevate this ideological position backed by natural science to a non-speculative level appropriate to its content, was not unproblematically fulfilled by the new natural science that achieved its breakthrough with Galileo and Kepler. So we have to investigate what these problems consisted in and what they are due to. In doing so, it is probably useful to first examine to what extent these problems can be traced back to the state of development of physics at that time.

4.2 Second Section

4.2.1 The Dualism Between Matter and Force

4.2.1.1 A Hint by Leibniz

In the text already mentioned above, *De ipsa natura sive de vi insita actionibusque creaturarum*, published in 1689, an argument with the Cartesian Sturm, Leibniz attacks Cartesian dualism as a physical theory worthy of criticism, although he uses metaphysical arguments. More precisely, he criticizes it in the form of the dualism between a purely passive, dead matter, characterized only by extension, and a force existing independently of it, ultimately founded in God. With this Leibniz himself gives the hint that the question has to be pursued in how far the mechanical physics in its content caused problems to the attempt to understand and to represent scientifically appearing reality as a coherence of inner necessity based on it. For it was not a whim of Descartes and his followers to regard force as something external to matter; rather, the relationship between matter and force represented one of the main problems of physics of the seventeenth century and beyond.

4.2.1.2 'Kinematic' Mechanics

Let us take a brief look at the state of this problem in seventeenth-century mechanical physics before Huygens and Newton. This period of development may be called the "kinematic" phase of mechanical physics, for which it is characteristic to concentrate on the study of the forms of motion, leaving aside the problem of force.[8] The dualism of matter and force thus seems justified by the state of physics before Huygens and Newton; the Cartesians seem to follow only 'kinematic' physics when they consider the action of forces, but not the forces themselves, explicable by a science investigating the material world. The connections are admittedly somewhat more entangled.

4.2.1.3 Galileo's Revision

If Galileo did not make any statements about the causes of the movements he investigated, i.e. about the forces, this resulted from a conscious self-imposed restriction. In this way, he tried to avoid unverifiable presuppositions playing into his investigation of the forms of movement, presuppositions about forces for whose scientific assessment the study of pure forms of movement would, he hoped, provide a viable basis. For him, the exclusion of the problem of forces was therefore a temporary abstraction based on a research strategy. However, Galileo's exclusion should by no means be understood as manifesting his view that the causes of motion, the forces, cannot be the subject of physics proceeding with mathematical methods.[9]

Which dynamic assumptions did Galileo exclude—of course he had a certain dynamic theory in the tradition of impetus physics—and which problems were contained in them? In his physics, at first glance, movements can be distinguished according to the Aristotelian scheme "natural vs. forced", from which a corresponding distinction can be concluded with regard to the causes of movement, i.e. the forces. Free fall and the so-called circular inertial motion peculiar to him can be called

[8] V gl. Dijksterhuis (1956, 360 pass); but cf. also the explanations deviating from this in detail in Laßwitz (1890, vol. 2, especially 3, 376 ff. and 470 ff.).

[9] Dijksterhuis (1956, 375 ff.).

"natural" motions, because they are inherent to corporeal things; all other motions, including obstacles and modifications of those "natural" motions, result from external effects of other bodies, and are thus to be understood as "forced" motions.

Of course, a significant difference to Aristotelian physics must be pointed out. Aristotle distinguished between "natural" and "forced" movements in the context of a cosmology, according to which a natural thing in its "natural" movements not only followed its own, isolated nature, but strived towards a cosmic place appropriate to its nature in relation to the nature of all other natural things. Thus, according to Aristotle, in its "natural" movement, a natural thing expressed a force in which its nature and the whole of nature were effective, through which the individual natural thing actuated its universal relationship to the cosmos and in which, conversely, the whole of nature thus appeared as a cosmic totality. This does not apply to Galileo's free fall or "circular inertial motion," even if it may seem to the superficial eye that Galileo was still very much attached to ancient cosmology because he imagined the universe to be spherical and finite. In Galileo's case, a body in the fall or in the "circular inertial motion" only follows its own, isolated nature, even if this is a general nature insofar as it is inherent in all bodies; the bodies thus express a general nature in the "natural" movements, but they do not relate to each other in this expression of force, thereby forming no cosmic connection.

There is no need to state here that this revision of Aristotelian physics must be seen in connection with the critique of the "substantial forms" of scholastic philosophy, from which effects and thus also movements are to be explained. More interesting is the result of this revision and the problems it contains. Even if Galileo still knows "natural" and "forced" movements in his implicit dynamics, in any case, due to that revision, the physical natural things stand next to each other as isolated, unrelated. Every change of movement that a body experiences through the action of another, is suffered as caused by a force external to it, regardless of whether the acting body followed a "natural" or a "forced" movement; and in the same way, the body itself acts on every other body only as a force external to it, regardless of whether it is in "natural" or "forced" movement. So whatever the cause of its movement, whether it comes from its nature or is "forced",—a moving body can only cause a "forced" movement.[10]

4.2.1.4 The Concept of Force of the Galileo School

In this respect, it was not inconsistent when Galileo's students, above all Baliani, in their attempt to formulate a uniform physical concept of force, interpreted all forces in the manner of causes for "forced" movements.[11] Methodologically, no other determination of the—as they are called today—force vectors could be considered than the one that starts from the ascertainable actions of force. On the one hand, it did not seem possible to distinguish from its effects whether the force action of a

[10] The relativity of the movement is left aside here.

[11] Cf. Dijksterhuis (1956, 400–409).

moving body was based on inherent forces or on a movement caused by the action of other bodies, i.e. was "forced". On the other hand, it was also not possible to see what this distinction was supposed to achieve. If this distinction thus proved to be irrelevant from the point of view of effect, and if at first a promising perspective seemed to open up for the scientific treatment of the causes of movement only if every expression of force, every movement was understood as an external movement caused by another body, then it also seemed that what Galileo's free fall and "circular inertial movement" implicitly contained—force as a capacity inherent in bodies—could be dropped. The Galileo school's interpretation of the causes of motion thus suggested a view of bodies according to which they were passively subject to external forces. Matter and force thus seem to be dualistically opposed to each other, while nature appears as a conglomerate of isolated bodies whose mutual effects on each other are external and accidental to them, even if the form of these effects has a law-like character.

4.2.1.5 Effects in the Corpuscular Theories

Against the background of the insight into the connection between matter and force that was gained at this stage of the development of physics, it becomes clear how difficult it was to try to understand and scientifically represent appearing reality as a coherence of inner necessity on the basis of this physics. These difficulties become vividly apparent in the corpuscular cosmologies of the seventeenth century. According to these theories, the universe represents a unity insofar as all its phenomena—according to the claim and under confident anticipation of later scientific verifications—can be traced back to structurally identical elementary components which, due to their structure, act on each other in a lawful form. Appearing reality is derived as a necessary result from these elementary building blocks and the lawful form of their influence on each other founded in their structure and in this respect is presented as a connection of inner necessity—except for the one point, which admittedly endangers the whole program, that only the form of their impacts on each other, but not also the cause of these impacts, is founded in the elementary building blocks.[12] This appearing reality requires force as an external principle, even if it is only a first impulse, and thus proves not to be founded in itself alone.

This problem of corpuscular theories can be studied in an exemplary manner in Cartesian cosmology, as found in the *Principles*, whereby it is sufficient here to refer to the distinction Descartes makes between the "general and original cause of motion" (*Princ.* II/ §36), the Creator God, and the "special causes" (ibid.) of motion, which act according to "unchangeable laws of nature" (II/§ 37). These "special causes" should not be understood as forces inherent in matter. Rather, they refer to the cause of the form in which the movement emanating from the Creator-God acts on the basis of the structure of the particles of matter according to certain rules

[12] Cf. Aristotle, *De Coel.* III 2, 300b8–16 (Cf. *On the Heavens,* 1984, 447–511).

derived from their geometric properties (II/§ 64). Matter itself has no inherent forces; the particles of matter can only pass on the movement originating in God in their interaction and thereby determine the form of the movement according to those rules.[13] Descartes, moreover, considers it necessary to assume that God constantly acts on the world, namely for the preservation of the quantum of motion (*Princ.* II/§ 36).[14] Thus the universe in force not only has a precondition external to it, but also requires the constant cooperation of a principle external to it for its reproduction or preservation. The program of grasping appearing reality as a coherence founded in itself and of presenting it scientifically can therefore, as much as it appears to be realized in the structural equality of the corpuscles and in the lawful mode of action of the forces based on it, be declared not to have been carried out, indeed not to be feasible on this basis. The fact that the universe, both with regard to the origin of movement and with regard to its preservation, does not appear to be founded in itself, this deficiency of corpuscular cosmologies of the seventeenth century reflects—as was to be demonstrated here—the state of development of mechanical physics before Huygens and Newton, its unsolved problems.

4.2.2 Is It Solely due to the State of Development of Physics?

On the other hand, it is important to ask why theorists such as Descartes generalized such contradictions in the temporary state of physics, such as the dualistic relationship between matter and force, in their cosmologies and elevated them to the status of a principle, instead of attempting—as Hobbes did with the "conatus"—to anticipate and speculate on possible solutions to the aporias contained in the state of physical research in the sense of the position that the reality that appears must be

[13] Cf. Dijksterhuis (1956, 461 f. and 464 f.); Laßwitz (1890, vol. 2, 118f.) ; but cf. also the interpretation deviating from this in the explanatory notes E. Cassirer (1998, 19 ff., see especially 24).—If one assumes that physicists of the seventeenth century, e.g. Galilei and Newton, understood inertia as a force, the fact that Descartes, in connection with that distinction, made his formulated his rectilinear law of inertia in connection with that distinction. Seems to contradict this interpretation. Descartes, however inertia as a force; with it Descartes only proves that all change of motion change of motion requires external influence (II/§ 37), but not that bodies that bodies resist a change of motion; cf. Laßwitz (1890, vol. 2, 62 f., 97 ff. and 105 f.)—Other corpuscular theorists in contrast to Descartes endowed the corpuscles with proper motion. Admittedly, two things have to be said about this. If for instance Gassendi attributes this proper motion to an impetus given to the matter particles by God (cf. Maier 1968, 29), then this can certainly not be regarded as a principally different solution of the problem of Matter/Force. Hobbes' assumption of what is called a "conatus" as "conatus", a constant, virtual being in motion of the matter particles. Can rather be seen as an approach to overcome the dualism between matter and force. and force (cf. Maier 1968, 45; Laßwitz, 1890, vol. 2, 216); but even on this basis appearing reality is not without contradictions as founded in itself, as will be shown later, when considering dynamic mechanics. Will show later, when considering the dynamic mechanics.

[14] Cf. Laßwitz (1890, vol. 2, 62). This problem will come back to us and become clearer in Newtonian dynamics. and become clearer.

understood as a coherence that is justified in itself. Descartes in particular suggests this question because the physical assumptions underlying his cosmology do not simply reflect the state of "kinematic" physics, but represent their own variant of "kinematic" mechanics that exacerbates the aporias. This is already indicated by the fact that Descartes—whose physical conception was apparently essentially mature and ready at the beginning of the 30s, i.e. before Galileo's "Discorsi" and before the elaboration of the concept of force by the Galileo school—elaborated the dualism between matter and force in contradiction to the state of physical research at the beginning of the 30s.[15] If this could perhaps be interpreted as an ingenious anticipation of later research by the Galileo school, it should be noted that the peculiarity of Cartesian physics, through which the aporias of "kinematic" mechanics are intensified and elevated to principles, cannot in any way be based on seventeenth-century physics. What is meant is the peculiarity that Descartes defines the elementary building blocks as structured solely by extension, but not also by something that would correspond to the later concept of mass. As little clarity as there was on this point in the physics of Descartes' time, it is nevertheless unmistakable on the other hand that the physicists of his time sought to get to grips with precisely the structural feature of material matter in terms of impenetrability, density, gravity, weight, etc., which the later concept of mass expresses and on whose formulation the overcoming of the dualism between matter and force inherent in "kinematic" mechanics depended.[16]

4.2.3 *Other Temporary Circumstances*

The reasons that specifically led Descartes to his physical conception are not as relevant in the context of this work as the reasons for the more general fact that corpuscular theoretical cosmologies in the Cartesian mould represent generalizations of "kinematic" mechanics that may well have points of reference in the latter, but cannot be explained in their contradictions from the problems of this mechanics alone. The direction in which the reasons for these generalizations are to be sought will only be briefly touched upon here, because in this chapter more important contradictions in the philosophical processing of classical physics in the seventeenth century, which concern the explanations of this work as a whole, must be mentioned.

[15] Cf. Laßwitz (1890, vol. 2, 117 ff.).

[16] Descartes also mentions "*soliditas*" as a determination of matter; but this means for him an intensional dimension of extension. Cf. *Princ.* III/§ 123; cf. Laßwitz (1890, vol. 2, 71 f.), 113 and. 97 ff. ; Maier (1968, 28).

4.2.3.1 The Mercantile Age

The one direction to be indicated here refers to the level of development of bourgeois society actually achieved in the seventeenth century. We can be brief about this with a clear conscience, because the theories of the seventeenth century have often been considered against the background of this stage of development.[17] It seems to me that the importance is not sufficiently appreciated of the otherwise common fact that the state of development of bourgeois society in the seventeenth century is characterized by a mercantilist economy—and this applies especially to post-revolutionary England—while only in the course of the eighteenth century, with the beginning of so-called liberalism, did the bourgeois mode of production achieve a breakthrough as a system reproducing itself by its own forces; at first, of course, essentially only in England.[18] Before the eighteenth century, therefore, the bourgeois mode of production and thus bourgeois society did not yet exist on its own. For the absolutist nations and territorial states of the seventeenth century, it is in any case true that they are based on a stage of development of bourgeois society on which it cannot exist and maintain itself independently. Two things can be inferred from England's state of development. On the one hand, the bourgeois mode of production was, under certain circumstances that cannot be discussed here, so far developed in the seventeenth century that it not only no longer needed feudalist agriculture, but also had to eliminate the feudalist mode of production, because, generally speaking, the identity, the core characteristic of feudalism, of political power and certain economic interests engaged in certain businesses and branches of the economy had become incompatible with the bourgeois mode of production at its present level.[19] But the fact that the state, even after the revolutionary elimination of feudalism, had to regulate economic events in a mercantilist manner is precisely how advanced England, on the other hand, presented bourgeois society as a system incapable of reproducing itself in the free play of economic forces, i.e. only obeying its own laws. Even at the high level of development of bourgeois society in seventeenth century England, bourgeois society required intervention from outside. With regard to our question as to why seventeenth-century corpuscular cosmologies developed the dualism between matter and force inherent in "kinematic" mechanics into a worldview according to which the universe requires an external, activating and regulating principle, this specific character of the bourgeois mode of production at that stage of development would certainly have to be given some weight.

[17] Reference is made here only to three widely used works: Borkenau (1934), Horkheimer (1930), Macpherson (1962).

[18] I use this term to describe the state economic regulations at the time of the transition of bourgeois society from the commercial-capitalist to the production-capitalist stage, i.e. the corresponding economic policy of both the absolutist and the constitutional states.

[19] Cf. Lefèvre (1971, 79).

4.2.3.2 The Flywheel

A second direction in which, in my opinion, answers to this question must be sought, refers to the development of technology in the seventeenth century, with regard to which we can also be very brief on the basis of the discussions in the first chapter of this work. We saw there ("A general bottleneck of technology") that the limited usability of natural driving forces on the traditional technical basis determined by craftsmanship grew into a bottleneck in the course of the seventeenth century, which the further development of technology as a whole depended on overcoming. It was not only a question of tapping into new natural sources of power, but also, and—as long as this process was not forthcoming—primarily a question of using the existing driving forces more economically, i.e. of making corresponding improvements to machine designs. If, under these circumstances, the "multiplication of forces" became the central program for the engineers of the seventeenth century, and the use of inertia through flywheels played an increasingly dominant role in its realization, a profound change took place in the conception of the means of work and thus implicitly of material matter. Until then, the machine had been viewed entirely through the lens of the craftsman's perception of his tool: as something that ideally transmits the external force moving it without friction losses, allowing it to act on the workpiece in an appropriate manner, i.e. as something passive. With this use of inertia, however, the working tool is seen as having inherent forces that must be released and made usable for production.[20]

[20] So that there is no misunderstanding: mechanical elements such as lever or pulleys have of course always been regarded as energy-saving; in this respect, the economical use of given forces by means of machines is not a peculiarity of the seventeenth century. With this use of mechanical The use of mechanical laws in this way could not justifiably give rise to the view that the force saved was added by the mechanism, i.e. that forces lay dormant in the mechanism as "potential energy". On the contrary, the mechanism was considered to be purely passive and even, with the development of ever larger machines, increasingly aggravated the problem that it consumed ever larger parts of the given forces unproductively, namely by friction.—In contrast, the flywheel designs of the seventeenth century attempted to utilize a force lying dormant as "potential energy" in the mechanism. The overcoming of the dualistic opposition of matter to an independent and separate force—the decisive prerequisite for the concept of energy—is becoming apparent, even though there was no real clarity about the connection between force and matter before the discovery of the law of conservation of energy by J. R. Mayer.—At this point it is perhaps appropriate to remark that in this work I use the term 'force' in a more general sense than in contemporary physics, not only because it is the terminus technicus used in this sense by the physicists of the seventeenth century; in my opinion it also offers the advantage that the difficulties which the history of physics has had with the problem of mechanical motion, and which have been obscured by the terms 'energy' and 'work', as well as the connection between theoretical advances and technical developments, are expressed more vividly in this term.

4.2.3.3 A Principled Circle of the Craftsman's Conception of Nature

This change in conception receives its first scientific expression from the dynamics of the second half of the seventeenth century, whereby the elaboration of dynamics by Huygens and Newton took place in close connection with this technical development.[21] This further development of physics at the same time throws light back on 'kinematic' mechanics and, beyond that, on impetus physics. On the one hand, it becomes clearer that the aporias of pre-dynamic mechanics are not least to be found in the artisanal form of production and the conception of nature contained in it, considered above (Nature as quarry), whereby it is interesting to see that the use of machines before the point of development considered here could not undermine this conception of nature.[22] At the same time, on the other hand, it becomes apparent that a principled circle arises if the position that appearing reality must be grasped as a self-grounded whole is based on this conception of nature. The general tendency of this conception of nature, founded in the artisanal form of production, to interpret 'natural' movements as 'forced' ones, provided, as the thrust against the 'substantial forms' of scholastic theory contained in it already showed, grist for the mill of that position by demystifying the universe and prepa ring the ground for its explanation based solely on appearing reality. The inevitable consequence of this tendency is, of course, that in this demystification it transforms the universe into a passive matter that requires God as an activating principle more urgently for its existence than the ancient cosmologies.[23] It can only serve the ideological position that appearing reality can endure without external preconditions by strikingly demonstrating God as a necessary external precondition in the consequence of its approach.[24]

[21] Cf. Wolff (1971), third part. Ms.; cf. also Bernal (1971, vol. 2, 457).

[22] The importance which the machine had for Descartes' thinking, shows this in an exemplary way; cf. Grossmann (2009/1935, 134ff/200 ff.).

[23] The remark of Hegel coined for the metaphysical systems of the seventeenth century thus also applies here (1896, vol. 3, 347–348) "God has the special privilege of having laid on Him the burden of what cannot be comprehended. The word of God is thus the makeshift which leads to a unity which itself is only hypothetical; for the process of the many out of this unity is not demonstrated. God plays therefore in the later philosophy a far greater part than in the early, because now the comprehension of the absolute opposition of thought and Being is the chief demand. [...] God is therefore, as it were, the waste channel into which all contradictions flow."

[24] For the impetus physics the tendency to interpret "natural" movement as "forced" is constitutive from the beginning, which must be shown above all by the fact that it is peculiar to it from Philoponus over Buridan and Oresme up to Copernicus and Galileo to understand the celestial motions, thus the most perfect 'natural' movements of the Aristotelian cosmology, as movements after kind of thrown bodies. The socio-historical circumstances of the formation of impetus physics in Byzantine Alexandria, its modified resumption in the Occident of the fourteenth century as well as its influence until exactly the middle of the seventeenth century give additional weight to the interpretation that it is a conception of nature corresponding to the artisan form of production. Cf. Wolff (1971).

Both the state of development of the bourgeois mode of production in the seventeenth century, in which it required interventions from outside to reproduce itself, whereby the increasingly cogent inherent laws of this mode of production made arbitrary interventions increasingly problematic, and the still predominantly artisan form of production with its image of a nature that is moved from outside, must be taken into account with regard to the reasons why seventeenth-century corpuscular cosmologies elevated the dualism between force and matter in "kinematic" physics to a principle in their view of the world.

4.2.4 Dynamics

4.2.4.1 Unsolved Problems

Of course, it must not be overlooked that the development of dynamic mechanics in the second half of the seventeenth century did not eliminate all the problems with which the attempt to understand appearing reality with the means of mechanical physics as a self-founded context was confronted. Just as, for example, the origin of the motion that modifies the gravity acting on the planets to the planetary orbit could not be explained by dynamic mechanics for the time being, so little could physics make statements about why universal gravity and the loss of motion due to friction had not long since brought all movements to a standstill and transformed the universe into a solidified lump of mass.[25]

[25] A passage in Newton's *Optics* expresses these problems so vividly that I cite it here at length: "Seeing therefore the variety of motion, which we find in the world, is always decreasing; there is a necessity of conserving and recruiting it by active principles: such as are the cause of gravity, by which planets and comets keep their motions in their orbs, and bodies acquire great motion in falling; and the cause of fermentation, by which the heart and blood of animals are kept in perpetual motion and heat; the inward parts of the earth are constantly warmed, and in some places grow very hot; bodies burn and shine; mountains take fire; the caverns of the earth are blown up; and the sun continues violently hot and lucid, and warms all things by his light. For we meet with very little motion in the world, besides what is owing to these active principles. And if it were not for these principles, the bodies of the earth, planets, comets, sun, and all things in them, would grow cold and freeze, and become inactive masses; and all putrefaction, generation, vegetation and life would cease, and the planets and comets would not remain in their orbs.

All these things being considered, it seems probable to me, that God in the beginning formed matter in solid, massy, hard, impenetrable, moveable particles; of such sizes and figures, and with such other properties, and in such proportion to space, as most conduced to the end for which he formed them; and that these primitive particles being solids, are incomparably harder than any Porous bodies compounded of them; even so very hard, as never to wear or break in pieces: no ordinary power being able to divide what God himself made One, in the first creation. While the particles continue entire, they may compose bodies of one and the same nature and texture in all ages: but should they wear away, or break in pieces, the nature of things, depending on them, would be changed. Water and earth, composed of old worn particles and fragments of particles, would not be of the same nature and texture now, with water and earth composed of entire particles, in the beginning. And therefore that nature may be lasting, the changes of corporeal things are

4.2.4.2 Theory of Interactions

But the overcoming of the dualism between matter and force, which began in Huygens' and Newton's dynamics, gave a decisive hint that and how appearing reality can be thought as a coherence founded in itself; and indeed dynamics gave this hint as a theory of interactions. That it is this character of dynamics which makes it appear not simply as a conclusion of previous mechanics, but as a completion in the sense of overcoming a main deficiency of previous mechanics, becomes clear above all in the third axiom of Newton's *Principia*—actio = reactio—prepared by Huygens' investigations of elastic impact, and in the theory of gravitation, which was controversial at the time. These theorems (whereby the third axiom presupposes the conception of inertia as a force) not only state that matter is not only subject to external forces but is itself endowed with force; according to them, all bodies are on the one hand in constant interaction (gravitation) and on the other hand can only act on each other as reciprocally acting forces (third axiom). The third axiom could still be understood—albeit only by abstracting essential moments—in the sense of the unifying structural equality of the corpuscular theories, as a theorem according to which the structural determinations of the elementary components such as extension, shape, density, etc. now also include, with the concept of mass, the determination of force, which is inherent to all bodies and co-determines the lawful form of their effects on each other. Universal gravitation, on the other hand, unambiguously establishes unity in a different way from the corpuscular theories; in it appears the dimension of being able to recognize appearing reality as a unity, as a whole, as a universe, because there is interaction among its parts, each part relates to all the others. In this way, dynamic mechanics as a theory of interactions at the same time reveals the core of the problems that arose from the dualism between matter and force in the attempt to comprehend and scientifically represent appearing reality as a coherence founded in itself.

---

to be placed only in the various separations, and new associations, and motions of these permanent particles; compound bodies being apt to break, not in the midst of solid particles, but where those particles are laid together, and only touch in a few points. It seems to me farther, that these particles have not only a *vis inertiae*, accompanied with such Passive laws of motion as naturally result from that force; but also that they are moved by certain Active principles, such as is that of gravity, and that which causes fermentation, and the cohesion of bodies" (Newton 1952, 400–401).

4.2.5 *The Paradox of Atomism*

The corpuscular theories expressed this problem in full clarity. In that the corpuscles configured themselves into appearing reality only due to external influences, this reality is not founded in the corpuscles as a necessity. It is true that the corpuscles, due to their structure, can only arrive at a certain configuration from a certain initial situation in a strictly deterministic way, and this configuration is thus necessary. But on the other hand, this necessity is not founded in the corpuscles, because they do not produce and reproduce their relations with each other on the basis of their own essential forces in such a way that the reality that appears would have its continuance in it, that its necessity would be founded in it. The often emphasized paradox of corpuscular theories consists in the fact that appearing reality must be understood in its necessity from the lawful interaction of such elementary entities, for which their interaction is not necessary but accidental.

On the basis of the dualism of matter and force, this posed an insoluble problem. But even the assumption, found in various forms among the atomists of antiquity as well as of the seventeenth century, that corpuscles are moved by themselves does not solve the core of the problem. Although this assumption succeeds in establishing the fact that the corpuscles act on each other in the corpuscles themselves, this assumption does not take anything away from the corpuscles' character of autarkic entities to which their interaction is external and accidental.

4.2.6 *Leibniz's Critique of Atomism*

In one of the passages where Leibniz critically deals with atomism, there is an almost classic formulation of this problem:

> In the beginning, when I had freed myself from the yoke of Aristotle, I accepted the void and atoms, for they best satisfy the imagination. But on recovering from that, after much reflection, I perceived that it is impossible to find *the principles of a true unity* in matter alone, or in what is only passive, since everything in it is only a collection or aggregation of parts to infinity.[26]

The critique of atomism is undoubtedly of constitutive importance for Leibniz's metaphysics; to understand the world or one of its beings not merely as a "collection ou amas de parties jusqu'a l'infini" is—negatively formulated—the program of his *Monadology*.[27] It is true that this bears unmistakable traits of the corpuscular theories, whereby formulations for the individual substances such as "atomes de substance" or "points metaphysiques" say less than the fact that in this metaphysics the unity of the whole is ultimately based, as in atomism, on the structural equality of the monads, which are equal to each other in their "qualites et actions internes"

[26] *Système nouveau de la nature* (Leibniz 1875–1890, vol. 4, 478; Leibniz 1989, 139).

[27] Cf. e.g. the fourth letter to Clarke, in: Leibniz, *Philosophische Schriften* (C.I. Gerhardt edition, Berlin 1875–1890), vol 7, 372.

determined by the concepts of "perceptions" and "appetitions".[28] But more interesting here than the question of whether Leibniz's metaphysics actually succeeded in overcoming the aporias of the corpuscular theories is the direction in which Leibniz attempts to overcome these aporias; for it is only in this direction that it becomes fully clear where Leibniz sees the core problem of the corpuscular theories. This direction is generally characterized by the thought that every monad relates to all other monads by virtue of its essence, that the connection of monads is therefore necessary because it is essential to every monad.[29] Their connection is—therein lies the point against the corpuscular theories—not a connection that is external to them, accidental, indifferent, but a connection that is essential to them, necessary. The external connection that nevertheless exists among them, and with regard to which physics comes into its own, is—certainly one of Leibniz's most powerful thoughts—founded in the structure of being-for-others, i.e. precisely as a necessary moment of the fact that the monads relate to one another essentially, on the basis of inner drive.[30]

4.2.7 Leibniz and Dynamics

Dynamics undoubtedly played a significant role in the development of this metaphysics. The definition of substance as "un Être capable d'Action" is not simply a philosophical postulate decreeing the overcoming of Cartesian dualism, as one might read Spinoza's solution; rather, it is most closely connected with Leibniz's concept of force developed in constant reception of dynamic mechanics, namely that of Huygens.[31] But Leibniz found valuable clues in dynamic mechanics not only with regard to overcoming the dualism between matter and force. In the interactions between the bodies explored by dynamics, the main deficiency of the corpuscular theories—namely that in the corpuscles as autarkic entities, the necessity of their interaction cannot be justified—was overcome entirely in the sense of the thought characterizing Leibniz's metaphysics; dynamic mechanics showed that the bodies stand in a necessary relationship to one another, namely one justified in their nature as masses.

If Leibniz did not regard dynamic mechanics as the basis for overcoming the aporia of corpuscular theory, despite the importance it had for his thinking, which can hardly be overestimated, but came to the conclusion that the means of physics are of no help here, but that metaphysical thinking is necessary in order to understand appearing reality as a coherence of inner necessity, this was not because he

[28] Cf. Simonovits (1968, 54 ff.); *Système nouveau de la nature*, Leibniz 1879, vol. 4, 482; *Principes de la nature* § 2, *Monadologie* §§ 14 and 15.

[29] *Principes de la nature* § 3, *Monadologie* §§ 11, 14, 15, 51, 52, 56.

[30] Cf. Feuerbach (1848, vol. 5, § 8, 60 pass.).

[31] *Principes de la nature* § 1; Cf. of the writings especially 'De primae philosoohiae emendatione' (1694), 'Specimen dynamicum' (1695), and the already mentioned *De ipsa natura sive de vi insita*; cf. also Feuerbach (1848, 46 f. and 76 f); Cassirer (1998, 314 ff); Holz (1958, 39 f.).

had misjudged the character of dynamic mechanics as a theory of interactions due to his rejection of Newtonian gravitation. It was precisely in this character that he saw its significance.[32]

4.2.8 *Leibniz's Critique of 'Blind Determinism'*

Now, in the context of this work, the metaphysician Leibniz would deserve no further interest if it were possible to assume that his theory stems from problems that have at best very indirectly to do with the ideological position of the rising bourgeoisie that appearing reality must be understood as a coherence of inner necessity. But this is not possible, as was already shown by the constitutive function that the critique of atomism had for the development of his metaphysics. With his metaphysics, Leibniz sought precisely to escape the contradictions that, in his view, inevitably arise when that ideological position attempts to base itself on the new science that has achieved a breakthrough in classical physics. One of the passages in which he counters the discrediting of teleology makes it clear and at the same time lets us see more precisely which contradictions it is that he hopes to escape with his metaphysics. "You know," says Theophilus to Philalethes, "that I used to go a little too far and had begun to incline to the side of the Spinozists, who […] disregarding the investigation of final causes, derive everything from a blind necessity."[33] Whoever takes appearing reality, as Leibniz could reproach not only Spinoza but also Hobbes or Locke for doing, as a connection of blind determination, has not understood it as a connection of inner necessity, and has therefore not redeemed the ideological position that appearing reality must be understood as a coherence of inner necessity, but has led it *ad absurdum*.[34] If, however, as Leibniz's rehabilitation of teleology suggests, "blind determinism" is an unavoidable consequence of the attempt to

---

[32] Cf. e.g. *De pr. ph. emendatione*, 1879, vol. 4, 470; *Système nouveau,* 1879, vol. 4, 484 ff.—The rejection of Newtonian gravitation could not have this meaning for two reasons. On the one hand, the dynamics already proves to be a theory of interactions in the theorem "actio = reactio"—Huygens' investigations of the elastic impact were in the center of Leibniz' occupation with dynamics; on the other hand, Leibniz' conception, developed in demarcation to the theory of absolute space (Guericke, Newton et al.), that space is a relation of real and possible bodies, (cf. above all the correspondence with Clarke, esp. the third and fifth letters (1879, vol. 7, 371 ff. and 389 ff.) together with the relativity theory of Huygens in the here decisive point on a mathematical level the same as the Newtonian gravitation, namely that the state of motion of a body is only scientifically ascertainable in its relation to all other bodies. Cf. e.g. *Système nouveau*, 1879, 486 f.

[33] *Nouveaux Essais sur l'entendement humain* I/I (1971, 36); "You know that I went a little too far elsewhere and that I began to lean towards the Spinosists, who […] despising the search for final causes derive everything from brute necessity" (1879, vol. 5, 65).

[34] Leibniz's accusation of a 'blind determinism' is certainly to be regarded as a polemical concept, which can be answered philosophically and historically, that Spinoza as well as Hobbes or Locke have to be judged more differentiated in this point. But, as so often, also here the polemical exaggeration seems to be better suited to draw attention to problems, than the all-considering view; for the context of this work, Leibniz's critique of 'blind determinism', as will be shown, puts the finger on the probably most contradictory point of the materialism of the time. For this reason I will use Leibniz's polemical term as a guideline for the further presentation.

understand appearing reality as a connection of inner necessity on the basis of the new science that has achieved a breakthrough in classical physics, then Leibniz's metaphysics is not only about avoiding this or that contradiction that results from this attempt. Rather, he thereby reveals that he sees this ideological position as fundamentally endangered by the new science. So it is no longer a question of individual deficiencies in the state of development of physics, which at the time made it difficult for that ideological position to rely on the new science.

4.2.9 *Third Question*

The contradiction in principle between the ideological position of the rising bourgeoisie and the new science that had made its breakthrough in classical physics, to which we had Leibniz point us, now deserves all our attention without question because it touches on the central point of the explanations of this book. If we have perhaps succeeded in showing that, and in what particular way, the development of bourgeois society made possible and drove forward the genesis of classical natural science, then this contradiction draws attention to the fact that the result of this genesis, the new science that emerged in the seventeenth century, forms a contradiction at least to the ideological position of the rising bourgeoisie; that is to say, this contradiction concerns the relationship between the genesis of classical natural science and the development of bourgeois society and leads us to expect that our previous explanations still have to be concretized by a decisive moment.

This concretization presupposes, of course, that the contradiction signaled by Leibniz's metaphysics is defined more precisely, that it is made clearer in what exactly the new science principally contradicts this ideological position. Leibniz's critique of "blind determinism" should serve as a guide. In this context, it is perhaps useful to first examine the question of whether and to what extent 'blind determinism' results from the new science in order to localize this principled contradiction more precisely.

4.3 Third Section

4.3.1 *A Tiny Difference*

The question of whether and to what extent this 'blind determinism' results from the new science that achieved a breakthrough in classical physics is perhaps easiest to answer by comparing the method underlying the corpuscular theories—in my opinion the most outstanding examples of what Leibniz criticizes as 'blind determinism'—with that of classical physics, limiting ourselves to a few basic features.

In this comparison, first of all, extensive agreement between the two methods becomes apparent. For there is no doubt that the analytical/synthetic method is constitutive for the corpuscular theories. The corpuscles and their properties are analytically developed and serve as starting points for the deductive reconstruction of phenomena.[35] The distinction between the so-called primary or objective and secondary or subjective qualities, which is characteristic of corpuscular theories, illustrates this methodological agreement particularly well.[36] This distinction is based on two closely related considerations. The secondary qualities are secondary because they can be deduced, namely as effects of the primary qualities that can be deductively reconstructed from them; they must be deduced because—unlike the primary qualities—they cannot be conceived as elements of a mathematically representable structure. At the same time, however, it becomes clear what the primary qualities are; they are the moments as which science when investigating natural interrelationships, must understand certain properties of natural things in order to be able to comprehend the general, lawful forms of these interrelationships (cf. above in Chap. 3, *Characteristic or moment?*). The properties of the corpuscles are now nothing other than these primary qualities. Appearing reality can be reconstructed from the corpuscles because their properties are the moments of lawful cause-effect relationships of appearing reality. In this respect, the corpuscular theories are without question a genuine and authentic expression of the method followed by the science that achieved its breakthrough in classical physics.

The difference between the method of the new science and that of the corpuscular theories can be seen in a seemingly tiny difference, namely in the fact that the corpuscular theories make the moments of the law-like interrelationships into properties of corpuscles, although they are in fact properties of natural things. If the corpuscles can thus be regarded as the very material existence of the initial concepts of a deductive system, this is certainly to be understood as an attempt to overcome the magical character that was inherent in the use of the constructive potency of geometry in the Renaissance (cf. above § *Application in technology*).[37] However, the fact that they secure the material character of the moments of lawful relations of action by conceiving of these moments as properties of corpuscles, rather than as properties of natural things, reveals a goal of knowledge of corpuscular theories that stands in an interesting contrast to that of the new science.[38]

---

[35] Cf. e.g. Descartes, *Princ.* II/ §§ 3 ff., 23 III/§ 1 et al.; the section‚Computatio sive logica' of Hobbes' *Elementorum philosophiae sectio prima—De Corpore* (Hobbes 1961) is perhaps the clearest presentation of these methodological foundations of corpuscular theories. Cf. for the rest Cassirer (1999, 38 ff. and 50 ff.); 1998, 19 ff.; Maier (1968, 26 ff., esp. 30.).

[36] For details, see above all Maier (1968).

[37] When Cassirer reproaches the corpuscular theorists for having misjudged, that the assumption of the objective existence of corpuscles is only a "logical postulate" (1999, 51), he for his part fails to recognize the qualitative difference in which the new difference in which the new physics stands to the thinking of the Renaissance. If this is consistent from his neo-Kantian position, he throws a light on the limits of the new he throws light on the limits of this position in the reconstruction of the history of science. The history of science.

[38] In order to avoid misunderstandings, it must be recalled that it is about the corpuscular theories of the seventeenth century, but not about modern atomic physics. With respect to the latter, it would, of course, also be wrong, because the structure of the atoms determines the properties of the natural things consisting of them, to overlook that the effects of the atoms, conversely, are also determined by the context which they form in a natural thing.—The corpuscular theories of the

If the moments of lawful interdependencies are properties of natural things, as corresponds to the method of classical physics, then the result of these interdependencies is necessary and accidental at the same time. It is necessary when natural things interact with certain properties between which, as moments of a general cause-effect relationship, a lawful cause-effect relationship exists; whether or not this is the case, however, is a fact independent of this general cause-effect relationship, presupposed by it and accidental; to this extent, the result of the lawful cause-effect relationship is also accidental, or, to put it differently, the result is only necessary under certain conditions. The knowledge of the lawful form of cause-effect relationships thus proves to be knowledge of the conditions under which certain results are to be achieved on the basis of the lawful form of a cause-effect relationship.

If, on the other hand, the moments of the lawful interrelationships are properties of corpuscles that are analytically deduced as the necessary conditions of lawful interrelationships in such a way that appearing reality must necessarily result from them according to the lawful form of the interrelationships, then the accent is not on the knowledge of the conditions under which certain results are to be achieved according to natural law, but on the fact that appearing reality as it is can be understood as the necessary result of an unchangeable nature.

It must therefore be stated that the 'blind determinism' criticized by Leibniz is not immanent to the new science, but results from the attempt, with the help of the new science, to present appearing reality as a coherence of inner necessity by tracing it back to an unchangeable nature. If, as we have seen, this attempt had to deviate methodically from the new science, this fact signals that the new science is not only not the cause of a deterministic interpretation of appearing reality, but rather contradicts such an interpretation. At the same time, however, it becomes clearer in which respect the new science, which has achieved a breakthrough in classical physics, contradicts in principle the ideological position that appearing reality must be understood as a coherence of inner necessity; it contradicts in principle the interpretative, contemplative component of this ideological position and the tendency contained therein to justify appearing reality as necessary.

---

seventeenth century, by the way, contain an ambiguity in this point due to the state of development of physics at that time. As far as appearing reality is conceived as a kind of composition of corpuscular building blocks, the difference, whether the moments of lawful we-contexts are conceived as properties of natural things or of corpuscles, seems to shrink to the fact that they are properties of a natural thing on the one hand and properties of the parts of a natural thing on the other hand. It is different, if the corpuscular theories think appearing reality—as in the case of the secondary qualities—not as a mere composition, but as an effect of the corpuscles in such a way that this effect appears as a new quality. If appearing reality, as the designation of the secondary qualities as subjective indicates, gets a character of mere appearance, this points to the fact that the corpuscular theories, due to this seemingly tiny methodological difference, tend to regard appearing reality as merely effected and itself incapable of any effect of its own.

4.3.2 The Endangered Perspective of Freedom

If the new science thus opposes an interpretation of appearing reality that understands it as a necessity to be accepted, it thereby proves to be true to that ideological position, namely its perspective of freedom; for it was precisely the refusal to understand appearing reality as a necessity to be accepted that this position was concerned with when it attempted to destroy the view that appearing reality was not founded in itself, but had external, supernatural and thus acceptable preconditions. The contradiction between the new science that achieved a breakthrough in classical physics and the ideological position that appearing reality must be understood as a coherence of inner necessity can thus be identified as the crisis-like intensification of a contradiction inherent in this ideological position itself; on the basis of the new science, the interpretative, contemplative component of this ideological position can only be redeemed by annihilating its perspective of freedom, in which the justification of Leibniz's fighting term 'blind determinism' comes to light.

4.3.3 What Is Leibniz Concerned With?

This is precisely what Leibniz actually had in mind; for it is undoubtedly constitutive of his critique of "blind determinism" to reject the view that something that can only be regarded as given must be recognized as necessary. This is what he is essentially concerned with when he defends the "*ideae innatae*" against Locke's *Essay concerning human understanding*. His concept of "windowless monads" must be seen in this context; it has to ensure that their representation of appearing reality cannot be given to the monads from outside, but comes solely from their inner spontaneity.[39] This intention of Leibniz's is perhaps most clearly shown in the fact that he grasps the nature of these spontaneous monads "*à l'imitation de la notion que nous avons des ames*".[40] For this has above all the sense that the relations in which the monads stand to each other and which establish the connection of appearing reality in its necessity must not be understood simply as a inevitable execution of their essential forces; these relations must rather have necessity *for* the monads; the coherence of appearing reality must be established in substances of subjective character in order not to be a blind coherence that has to be accepted, but one of inner necessity. If the ideological position, which understands appearing reality as a coherence founded in itself, was directed against the privileges, claims to power and paternalism invoking an otherworldly being, then it only appears as a consequence of this position when in Leibniz's philosophy the subject does not allow itself to be dictated what is necessary, but emerges as the instance before which what must be considered necessary must prove itself. In this respect, there is no question that his

[39] Cf. Horn (1965, 50 f).

[40] *Système nouveau,* 1875–90, vol.2, 479, *Monad.* § 19.

critique of 'blind determinism' is committed to the freedom perspective of that ideological position.

If Leibniz, in the critique of Spinozism as well as of the deterministic implications of Lockean philosophy, appears as a zealous defender of religion, opposes precisely the atheistic consequences of the position that appearing reality must be understood as a coherence justified in itself, then, in view of 'blind determinism', it must certainly be considered that it is indeed less offensive to a rational being to recognize a world created by God under salvific intentions as necessary than a blind nature.[41] But at the same time, it becomes clearer why Leibniz, in his critique of 'blind determinism', did not fall back on the critical potential that—as seen—was contained in the new science itself against 'blind determinism', but on the contrary saw 'blind determinism' as founded in the attempt to redeem that worldview position on the basis of the new science. This is consistent because he is no less committed to the contemplative component of this position than the criticized determinists. So Leibniz sees not only the freedom perspective of this position, but also its contemplative component endangered by 'blind determinism'; but the latter is endangered precisely because of the critical potential of the new science, which comes to light in the fact that it takes revenge on a contemplative use that is inappropriate to it, in that such an interpretation of the world on its basis is only able to grasp the necessity of appearing reality as a blind determinative context, i.e. precisely not as an inner necessity.

4.3.4 'Substance as Subject'

That Leibniz was not concerned with a critique of the contemplative use of the new science in his critique of 'blind determinism', but with a different interpretation of the world, one more appropriate to the subject, is shown by his metaphysics, which can be ascribed to the program of taking 'substance as subject' (Hegel) as can the philosophy of Spinoza and Descartes, whose "cogito ergo sum" would lose its specific meaning if thinking did not assure itself of its own substantiality and thus at the same time of the subject character of substance, even if Descartes, unlike Spinoza and Leibniz, did not see how it could be avoided that another, unconnected blind substance faced this subjective substance.[42] The contemplative character emerges not least in the receptivity and passivity of the subjective substance criticized by Hegel, which Leibniz's efforts around the concept of "vis activa" were not able to change decisively; the subjective substance "does not generate" its ideas "in itself" (Hegel), they are rather innate to it, or rather, the subjective substance does indeed

[41] Cf. e.g. *Considérations sur la doctrine d'un esprit universel unique*, 1875–90, vol. 6, 336 f; see also the preface of the *Nouveaux Essais* (Leibniz 1971).

[42] Hegel (1977, 10); cf. Rohs (1969, 24 f.).

bring forth its ideas, but not "as a force generating and uniting them", rather "arise in the monad like bubbles" (Hegel).[43] The contemplative character of these metaphysics committed to the program of taking the "substance as subject" appears in its principal form, however, in the epistemological dimension they received from Descartes' *Meditations*.

4.3.5 Necessary Truth

Leibniz's distinction between "*vérités nécessaires ou de raison*" and "*vérités contingentes ou de fait*" is not primarily concerned with the epistemological question of which truth is more certain in the sense of excluded error; the "*vérités de fait*" are truths and have—sources of possible error taken into account and eliminated—no less certainty in this narrow epistemological sense than the "*vérités de raison*".[44] The purpose of this distinction can be seen in Leibniz's letter to Bierling of 19.11.1709: "If he (Locke) had correctly considered the difference between the necessary truths, or those which are known by demonstration, and those which we arrive at to a certain extent by induction, he would have seen that the necessary truths can only be proved from the principles implanted in the mind, the so-called innate ideas. because the senses teach what happens, but not what necessarily happens."[45] It is thus a question of distinguishing between truths that are of an actual nature and are only to be accepted as given, and truths that are to be recognized as necessary. How this necessity must be justified can perhaps be better illustrated by Descartes than by Leibniz.[46]

---

[43] Cf. especially *De pr. ph. emendatione*; Hegel (1969, 396).

[44] Cf. e.g. *Monad.* §.§. 33, *Nouveaux Essais* I/1 (1971, 38 ff.), IV/2 (419 ff.); Cf. e.g. *Nouveaux Essais* IV/2 (437 f.), IV/6 (480); *Discours sur la métaphysique* § 13.

[45] "*Si discrimen inter veritates necessarias seu dernonstratione perceptas, et easquae nobis sola inductione utcunque innotes sunt, satis considerasset, animadvert isset necessarias non posse cornprobari nisi ex principiis menti insitis,cum sensu quidem doceant, quid fiat, sed non quid necessario fiat.*" (1879, Vol. 7, 488); that Leibniz here defends the '*ideae innatae*' exactly in the sense of Descartes, see *Princ.* I/§ 49.

[46] If we do not regard the Cartesian doubt as the foundation of an epistemological position, but as a founding step of a new metaphysics, then this has some plausibility because of the perspective from which the philosophy of the seventeenth century has been considered. There is, of course, a long-running interpretive controversy on this question, which should at least be pointed out here, if we cannot go into it. A short but instructive overview of this controversy can be found e.g. in Fahrenbach (1966, 64–91).

4.3.6 *Cartesian Doubt*

The fact that Descartes refers back to the ontological proof of God at a crucial point in the *Meditationes*, which is generally criticized as circular or perceived as embarrassing, seems to me to be revealing in our context, if we remember Descartes' reasons. It would undoubtedly be too short-sighted to think that Descartes assures himself of the existence of God so that the "*ego cogitans*", referred to itself by radical doubt, can find its way out of the isolation of solipsism and be assured that something objective, something that exists outside him, corresponds to his ideas. Neither the question of whether something exists outside the "*ego cogitans*", nor the question of whether the subject can recognize something that exists outside it, underlies Cartesian doubt, nor, finally, whether the subject can recognize this reality correctly; for Descartes, the reality that appears cannot be of a different nature than that which the intellect, proceeding according to the correct method, grasps.[47] The doubt has its basis rather in the fact that the subject is not able to establish the existence of this reality as necessary as its essence.[48]

4.3.7 *Necessary Existence*

This fact is due not primarily to the subject, but to this appearing reality itself: Its existence does not necessarily follow from its essence.[49] This is the fundamental problem that Descartes' doubt reflects. Once the question is posed in this way, it would only be an escape from the consequences if Descartes had tried to solve this problem by recourse to an innermost, logical, transcendental or whatever structure of the "*ego cogitans*"; the "*ego cogitans*" shares with appearing reality that in its essence its existence is not founded as necessary.[50] The recourse to the ontological proof of God is thus by no means inconsistent within this question. Something in whose essence its existence would at the same time be founded as necessary would fulfill the demands that are made in order to be able to think of it as a reality of inner necessity. Leibniz, who—as already mentioned—criticized Descartes' use of the ontological proof of God, aims at the same solution as Descartes by means of the cosmological proof of God; he too is concerned with a reality that is the "*raison*

---

[47] Cf. e.g. *Regulae* XII/ §§ 5 f., 14 f., *Princ.* II/1, op. cit.; Cf. e.g. *Medit.* V/§§. 9–10.

[48] Cf. ibid., and also Descartes' replies to the Objections to the *Meditations*, and his letters to Mersenne of 27. 5.1630 and letter to Mesland of 2.5.1644 (AT I, 151ff., AT IV, 111ff./CSMK 25, 231).

Fahrenbach's term "factual evidence" is meant, if I understand him correctly, to express just this (1966, 83).

[49] Cf. I/§§. 15–16 and *Medit.* III, esp. § 33. Cf. also Halbfass (1968, 124 ff. esp. 133).

[50] Therefore, in the third meditation a second proof for the existence of God is necessary.

suffisante" of itself.[51] Only in such a reality of inner necessity could appearing reality be justified as necessary.

4.3.8 'The Best of All Possible Worlds'

If we disregard the objections that may be raised in detail against this solution on the part of a philosophy that is itself metaphysical, it is in any case clear that a given reality that a rational being should be able to recognize as necessary must in principle be thought of in this way. The subject, in this respect this metaphysics differs from earlier ones, certainly has a central position in this; for it is the criteria founded in its reason by which the necessity of reality is measured.[52] However, by clarifying in this metaphysical train of thought how it can think appearing reality as a coherence of inner necessity, the subject only clarifies the conditions under which it can recognize reality as it is as necessary, under which it can ascribe the character of necessity to given things. If the result of this clarification is not 'blind determinism', it is nevertheless a theory which, by grounding appearing reality in an existence whose existence necessarily follows from its essence, makes appearing reality to be accepted in a way that far surpasses any conceivable 'blind determinism'. If a deterministic theory explains appearing reality as the necessary result of an unchanging nature, it thereby simultaneously prohibits itself from that justification of appearing reality towards which that metaphysics tends, the justification of reality as the 'best of all possible worlds'. In this respect, Hobbes, for example, proves himself to be a seventeenth-century Voltaire when he only admits as an object of science that which is produced, i.e. that which does not have the reason for its existence in itself.[53] Against this backdrop, 'blind determinism' itself proves to be a form of remaining faithful to the freedom perspective of the ideological position of the rising bourgeoisie, a form, admittedly, whose obvious contradictoriness we will deal with later.

The criticism of 'blind determinism' was therefore not directed at the contemplative use made of the new science. On the contrary, it was directed at the threat to the contemplative character of the ideological position that appearing reality must be understood as a coherence of inner necessity by the new science that had achieved a breakthrough in classical physics, a threat to which 'determinism' in Leibniz's

---

[51] Cf. e.g. *Monad.* §§. 36–38.

[52] In the elaboration of the character of Leibniz's metaphysics, which stands out from the tradition I see a great strength of the most criticized Leibniz book by H.H. Holz (1958) quoted above (cf. esp. the second chapter).

[53] Cf. *Element. ph. sect. pr.—De Corpore, Pars prima* I/8 (Hobbes 1961).—Of course, this is not to be understood as if Hobbes would recognize as the object of science only what can be generated by the head in the manner of mathematical derivations, and not also something whose thought-independent generation can only be comprehended by the head. Hobbes is concerned with a demarcation between science and theology; a being whose existence necessarily follows from its essence is regarded by him as the object of theology, not of science.

eyes had already secretly succumbed when it presented appearing reality as a blind connection of determination, i.e. not as a connection of inner necessity. This critique of 'blind determinism', whatever may be said about its premises, certainly hits the mark and at the same time confirms our explanations in the third chapter. In fact, on the basis of the new science, the reality that appears cannot be justified as necessary; this does not even succeed if methodological corrections are made to it in the manner of the corpuscular theorists.

4.3.9 The New Science and the Absolute

The new science, which attempts to explain every appearance by deriving it from its lawful conditions of origin, does indeed—as already explained—conceive of appearing reality as it is, as necessary insofar as every appearance results necessarily from the respective given circumstances according to the laws inherent in appearing reality. But it is peculiar to it to be able to understand appearing reality in this way only as necessary, in that it shows at the same time that every appearance is necessary only under those respectively given conditions, thus always only relatively necessary. It can only explain the necessity of an appearance by indirectly showing the conditions under which it would not be necessary. Thus, if it decodes appearing reality as a necessary reality by finding out the possibilities of its otherness, of its change, which lie in appearing reality, then it simultaneously honors the ideological position that appearing reality must be understood as a connection of inner necessity—only not in a contemplative way. It shows that appearing reality is an absolute, because it can only be determined by its own laws and on the basis of the possibilities it contains. But this does not negate man's self-determination, which was the concern of that ideological position in its perspective of freedom, but concretizes it; man can only change appearing reality according to his necessities if he seizes and uses the possibilities and laws contained in it. appearing reality is—this is what the new science points out in contrast to a contemplative theory—necessary as a precondition of its determination and change. Bacon's motto that we must obey nature in order to master it epigrammatically sums up this "ideological" content of the new science.

4.3.10 The New Science and Contemplative Theory

Without doubt, this constitutively practical character of the new science left deep traces in its processing by the philosophical systems of the seventeenth century, whereby it is probably not surprising that these traces are particularly evident in the metaphysical systems of Descartes and Leibniz; their program of grasping "substance as subject" can be described—as paradoxical as it may sound—as a transposition of this practical character to a contemplative level.[54] The constructive character of deterministic theories is, of course, no less a clear reflection of the constitutively practical character of the new science that made its breakthrough in classical physics. However, if we had to state that deterministic materialists as well as metaphysicians represented the position that appearing reality must be understood as a coherence of inner necessity in the sense of a contemplative interpretation of the world, this means that both directions were just as undoubtedly unable to take up and use that constitutively practical character of the new science for that ideological position. The examination of Leibniz's critique of "blind determinism" thus reveals to us a principled contradiction between a constitutively practical science that emerged in classical physics and an equally constitutively contemplative philosophical processing of that science on the part of deterministic materialists and metaphysicians alike.

If the philosophical processing of the new science that made its breakthrough in classical physics considered here was unable to make its specificity, its constitutively practical character, its own, then this fact is of great importance for the explanations of this book. For this fact, as has perhaps already become clear, must inevitably have repercussions on the attempt to provide for the position that appearing reality must be understood as a connection of inner necessity, which for so long could only be held speculatively, a non-speculative, precisely a scientific basis with the help of the new science, in order to be able to turn it—thus in a qualitatively new way—against the feudalistic world view. These repercussions can be seen in Leibniz in an extreme form. When he—unlike the materialists—draws the conclusion from the fact that appearing reality cannot be justified as necessary with the new science, that that worldview position precisely cannot be redeemed experientially, then this means that he considers the retention of the speculative form of that worldview position to be indispensable for the salvation of its contemplative character, that he thus rejects precisely the overcoming of its speculative form. On the other hand, these repercussions are equally evident in the deterministic materialists, when they can only strip that ideological position of its speculative form with the help of the new science, by leading it ad absurdum not only in its contemplative character, but at the same time also in its perspective of freedom.

This points us to the fact, which is important for the overall explanations of this book, that with the breakthrough of the new science in classical physics, by no means all conditions were already given for stripping the position that appearing

---

[54] With regard to the explanations of the third chapter, the following 'trace' deserves special attention: in Descartes and Leibniz are found the clearest reflections of the moment characterizing the new science, that their concepts and propositions are directed to the representation of relations and conditions and not to those of things and their properties. Cf. for Descartes e.g. *Discours* II/§ 11 and for Leibniz e.g. *Princ. de la Nature* §. 5 and *Monad.* §§. 88; cf. *Nouveaux Essais* II/25. Also the assumption of the innate ideas reflects not least this fact; cf. for Descartes e.g. *Med.* V/§ 11 and *Princ.* I/§ 49 and for Leibniz *Nouveaux Essais* I/1 (1971, 46); cf. also Heimsoeth (1914, 40 ff.). This is the qualitative difference of the new science compared to the knowledge of an experienced craftsman or engineer, which Leibniz expresses directly in this context: *Nouveaux Essais* I/1 (1971, 53 f).

reality must be understood as a coherence of inner necessity of its speculative form at all, or without producing new contradictions. The reasons for the difficulties with which the attempt to place this ideological position on a scientific basis is obviously still confronted are—this is shown to us at the same time—none other than those that are to be held responsible for the inability to take up the specificity of the new science, its constitutively practical character, to make this character the specificity of theory beyond the natural sciences.

4.3.11 Determinism and the Perspective of Freedom

The reasons for the constitutively contemplative character of the philosophical systems considered here, to which we must therefore turn, are perhaps most easily revealed by the aforementioned contradictory form in which even blind determinism proves to be committed to the freedom perspective of that ideological position of the rising bourgeoisie. If this was evident in a general form in the resistance of deterministic materialism to the tendency to justify the existing reality, it is shown concretely in the practical role that deterministic materialism played as a theory of the bourgeois revolution from Hobbes to the so-called ideologists of the great French Revolution and beyond.

The potential for attack that deterministic materialism contained in its atheistic consequences against the Church, one of the main cornerstones of feudalism, is less important in this context than the way in which deterministic materialism was to be directed as a weapon against absolutism. As is well known, it received this thrust above all as the foundation of a theory of the state that assumed unchangeable, quasi-natural foundations of social coexistence. Hobbes' social contract still represented the step by which the human race left the natural state of the *bellum omnium contra omnes* by constituting both state and society at the same time; at the end of the seventeenth century, when in England not only absolutism had been overcome, but also the dissolution of the mercantilist phase of bourgeois development was in the offing, Locke's theory of society and the state assumed that the society of private property was a not constituted, but quasi-natural order of human coexistence that preceded and underlay all constitution of the state and whose autonomy every government had to respect. Since, according to this theory, society is not determined by the laws of politics, but vice versa, politics is determined by the laws of society, it not only denies absolutism all justification, but also convicts it of being an intolerable disturbance of a natural order. The contradictory form in which deterministic materialism is committed to the freedom perspective of the ideological position of the rising bourgeoisie reveals its logic here; if the theories of state and society based on deterministic materialism show the disastrous consequences of the aristocracy not submitting to the laws of that quasi-natural social order, then they are at the same time saying that the freedom of the individual can only be secured if all members of society submit to bourgeois social relations as a quasi-natural order.

If bourgeois society opposes feudal society in this way as a kind of natural state of the human race and thus condemns feudal society as a dangerous degeneration and decadence, as something unnatural, then this is—as explained in the second chapter—neither to be seen as a merely brilliant propagandistic maneuver in the struggle against absolutism, nor as an apologetic move to cover up the contradictions inherent in the bourgeois mode of production. Rather, these theories of state and society give expression to the objective fact that the bourgeois social relations could be understood—here lies the difference with the social formation under the dictatorship of big landownership—as an earthly coherence founded in itself, but not as relations that can be controlled by people.

4.3.12 *Two 'Natures': Two Theories*

If these deterministic materialist theories of state and society prove to be a realistic expression of the fact that the social relations determined by the bourgeois mode of production determine individuals like nature, which is not under control, then their often elaborated character of being the expression of a "necessary appearance" (Marx) comes to the fore in the fact that they present bourgeois social relations as a kind of natural order of the human race and thereby—this is the decisive point for our context—conceive the uncontrollability of social relations as an unchangeable law of nature. Nature, as bourgeois society conceived in this way presents itself, is thus principally distinguished from ordinary nature by the fact that it is regarded as uncontrollable. Bacon's motto that we must obey nature in order to control it is precisely not applicable to the nature as which bourgeois society appears in the deterministic materialist theories of state and society; rather, this nature must be left in its own autonomous course; any intervention in the course of this nature would have catastrophic consequences.[55]

The character of the uncontrollability of this nature now inevitably determines the mode of its cognition. A cognition that examines its laws in order to know under what conditions other social conditions would be possible is not appropriate to bourgeois society understood in this way; rather, a cognition that understands it as the necessary result of unchangeable principles corresponds to it. For the knowledge of this nature, the form of theory that has made its breakthrough in classical physics is not appropriate, precisely because of its constitutively practical character.

[55] F. Borkenau in his Descartes analysis (1934, 268 pass.), which differs thoroughly from our expositions, comes to a corresponding result: "The goals of bourgeois philosophy: mastery of nature through insight into natural events, on the other hand, insight into social events, merely in order to take it as is" (1934, 301).

4.3.13 The Anticipatory Moment of Modern Natural Science

The deterministic materialist theories of state and society are a particular form of the view that the social nature of man is something immutable., which presupposes the high level of development of bourgeois society in some European countries at the time of the seventeenth century. This view itself becomes recognizable as the premise which lends stringency to the contemplative character of the philosophical treatments of classical physics considered in this chapter and to their characteristic inability to take up the specificity of the new science that had made its breakthrough in classical physics, namely its constitutively practical character, and to make it the specificity of theory beyond the natural sciences. If this conception, for its part, was based on the actual uncontrollability of the social relations determined by the bourgeois mode of production, then at the same time the conditions became apparent under which the form of theory that emerged in classical physics could be universalized; its constitutively practical character can become the essential feature of theory in general if the conditions can be changed on which the uncontrollability of social relations was based for so long. The principal contradiction in the philosophical processing of classical physics, which we dealt with in this chapter, can thus, as it seems, in the last instance be traced back to the fact that the qualitatively new, virtually unlimited power of man over nature, the gain of which modern natural science stands for, could not at that time be accompanied by a corresponding power vis-à-vis man's own, historical-social nature.

If the theoretical form appearing in classical physics thus points beyond the limits of bourgeois society, we saw in the explanations dealing with the preconditions and conditions of the genesis of this theory that the new science, together with its anticipatory moment, could nevertheless have the development of bourgeois society as its basis; however, it is perhaps only from the recognition of this anticipatory moment that the decisive significance of the circumstance can be appreciated that this theory developed its scientific form—as explained in the third chapter—in the closest connection with material production. The contradiction that the speculative form represented for bourgeois materialism is confirmed as the driving moment of this genesis by the explanations of this fourth chapter, but in such a way that at the same time it became clear that the breakthrough of the new science in classical physics was not in itself able to eliminate the contradictions of this materialism. If, with the help of the new science, materialism was able to develop into an experientially oriented theory and thus counter the speculative theory of feudalism in a qualitatively new way, on the other hand it still agreed with this speculative theory in that it "only interprets the world differently" (Marx).[56]

[56] Marx, 11th thesis on Feuerbach, MECW 5, 8.

4.3.14 The Way Forward

If the principal contradiction between this contemplative theory and the new, constitutively practical science appearing in classical physics shows once more how mistaken it would be to regard modern natural science as an ideology founded in the conditions of bourgeois society, it would, on the other hand, be interesting to examine whether and how the conception of nature contained in that premise of contemplative theory, i.e. in the conception of unchanging social nature, has a retroactive effect on the self-understanding of the natural sciences and perhaps even impairs their knowledge.[57] In my opinion, the insight to be gained from the reconstruction of the genesis of modern natural science, that with this modern natural science a form of theory appeared that had the development of bourgeois society as its precondition and at the same time points beyond it, should form the starting point of such an investigation.

References

Aristotle. 1984. *The complete works of Aristotle*, ed. J. Barnes, 2 vols. Princeton: Princeton University Press.

Bernal, J. D. 1971. *Science in history*. 4 vols. Cambridge MA: MIT Press (First edition London 1954).

Borkenau, F. 1934. *Der Übergang vom feudalen zum bürgerlichen Weltbild*, Schriften des Instituts für Sozialforschung. Vol. 4. Paris: Alcan. (Repr. in Borkenau/Grossmann, Darmstadt: WBG 1971).

Cassirer, E. 1998. *Leibniz' System in seinen wissenschaftlichen Grundlagen*, Gesammelte Werke. Vol. 1. Hamburg: Felix Meiner Verlag. (first ed. Marburg: Elwert 1902).

Cassirer, E. 1999. *Das Erkenntnisproblem in der Philosophie und Wissenschaft der neueren Zeit. Zweiter Band*, Gesammelte Werke. Vol. 3. Hamburg: Felix Meiner Verlag. (first ed. Berlin: B. Cassirer 1907).

Dijksterhuis, E. J. 1956. *Die Mechanisierung des Weltbildes*. Berlin-Göttingen-Heidelberg.

Fahrenbach, H. 1966. Endlichkeit des Bewußtseins und absolute Gewißheit bei Descartes. In *Metaphysik und Subjektivität. Festschrift für W. Cramer*, 64–91. Frankfurt am Main: Klostermann.

Feuerbach, L. 1848. *Sämtliche Werke*. Vol. 5. Leipzig: Wigand.

Grossmann, H. 2009. The social foundations of the mechanistic philosophy and manufacture (1935). In *The social and economic roots of the scientific revolution (BSPS 278)*, ed. G. Freudenthal and P. McLaughlin, 103–156. Berlin: Springer.

Halbfass, W. 1968. *Descartes' Frage nach der Existenz der Welt: Untersuchungen über die cartesianische Denkpraxis und Metaphysik*. Meisenheim am Glan: Hain.

Hegel, G. W. F. 1896. *Hegel's lectures on the history of philosophy*, transl. by E. S. Haldane and F. A. Simson. 3 vols. London: Routledge.

[57] In my essay on Second Nature and the Natural Sciences (Appendix B in this volume) I have pointed out some aspects of such an investigation which seem essential to me.

Hegel, G. W. F. 1969. *Hegel's science of logic*, trans. A. V. Miller. London: Allen & Unwin.
Hegel, G. W. F. 1977. *Phenomenology of spirit*, trans. A. V. Miller. Oxford: Clarendon Press.
Heimsoeth, H. 1914. *Die Methode der Erkenntnis bei Descartes und Leibniz* (1. Part). Giessen: A. Töpelmann.
Hobbes, T. 1961. *Elementorum philosophiae sectio prima de corpore*. Aalen: Scientia.
Holz, H. H. 1958. *Leibniz*. Stuttgart: Kohlhammer.
Horkheimer, M. 1930. *Anfänge der bürgerlichen Geschichtsphilosophie*. Stuttgart.
Horn, J.Ch. 1965. *Monade und Begriff*. Wien-München.
König, J. 1946. Das System von Leibniz. In J. König, *Vorträge und Aufsätze*, ed. G. Patzig (pp. 27–61). Freiburg: Alber.
Laßwitz, K. 1890. *Geschichte der Atomistik*. 2 vols. Hamburg: Leopold Voss.
Lefèvre, W. 1971. *Zum historischen Charakter und zur historischen Funktion der Methode bürgerlicher Soziologie*. Frankfurt am Main: Suhrkamp.
Leibniz, G. W. 1875–1890. *Philosophische Schriften* (C. J. Gerhardt ed.). 7 vols. Berlin: Weidmann.
Leibniz, G. W. 1879. *Philosophische Schriften* (G.J. Gerhardt ed.). Berlin.
Leibniz, G. W. 1971. *Nouveaux Essais sur l'entendement humain (ed. E. Cassirer): Neue Abhandlungen über den menschlichen Verstand*. Hamburg: Meiner.
Leibniz, G. W. 1989. *Philosophical essays*, trans. R. Ariew & D. Garber. Indianapolis: Hackett.
Macpherson, C. B. 1962. *The political theory of possesive individualism – Hobbes to Locke*. Oxford.
Maier, A. 1968. *Zwei Untersuchungen zur nachscholastischen Philosophie*. Rome: Edizioni di Storia e letteratura.
Newton, I. 1713. *Philosophiae naturalis principia mathematica*, 2nd, ed. R. Cotes. Cambridge: Cornelius Crownfield.
Newton, I. 1952. *Opticks*, 4th (1730), ed. I. Bernard Cohen et al. New York: Dover.
Rohs, P. 1969. *Form und Grund*. Bonn: Bouvier.
Simonovits, A. 1968. *Dialektisches Denken in der Philosophie von G. W. Leibniz*. Berlin: Akademie-Verlag.
Tomberg, F. 1973. *Bürgerliche Wissenschaft*. Frankfurt am Main: Fischer Taschenbuch.
Wolff, M. 1971. *Fallgesetz und Massebegriff: Zwei wissenschaftshistorische Untersuchungen zur Kosmologie des Johannes Philoponus*. Berlin: de Gruyter.

Appendices

Appendix A

Second Nature and the Natural Sciences: Reflections on the Tasks of Philosophy in the Discussion About the Foundations of the Natural Sciences
Wolfgang Lefèvre
Lecture, June 1975 at the University of Konstanz. (*Philosophia Naturalis* XVII(2) (1978), 242–253).

The upheaval in the physical worldview at the turn of the last century seems to have lent the discussion about the foundations of the natural sciences a significance that it has not had since the breakthrough of the classical natural sciences in the seventh century.[1] Of course, this discussion never ceased in the eighth and ninth centuries, and it would certainly not be easy to reject from the outset that the discussion of foundations in our century falls short of the level of the Kantian or Hegelian

[1] A useful introductory bibliography on contemporary philosophy of science can be found in: Krüger (Ed.) 1970, 509 ff. On the philosophy of science position of the so-called Erlanger Schule, cf. above all Kambartel and Mittelstraß (eds.) 1973, and the bibliography in Janich et al. (1974). Of the literature available on this subject in Eastern European countries, at least a few easily accessible books should be mentioned: the proceedings of the 1958 Moscow All-Union Conference on the Philosophical Questions of Natural Science (translation, Kosing [for Author Collective] 1968); Harig and Schleifstein (eds.) 1960; Griese and Laitko (eds.) 1969; Hörz 1976.

On the subject of science and society, there is a good short bibliography at the end of the introduction to Weingart (Ed.) 1972, and at the end of the introduction to Diederich (Ed.) 1974. Of the literature available on this subject in the Eastern European countries, reference should again be made to at least a few easily accessible books: Kosel (1957), Harig and Neels (Eds.) (1963), Gündel (ed.) (1968), Wolkow (1970).

For the reflection of the problems addressed in the following by the natural scientists themselves, reference is made only to the following examples: Born (1965), Planck (1946/1949), Russell (1931), von Weizsäcker (1957/2011).

W. Lefèvre, *Theory of Nature and Mode of Production*, Political Epistemology,
https://doi.org/10.1007/978-3-032-02295-0

discussion of the foundations of the natural sciences. However, it seems to me that the foundational discussion of the eighteenth and nineteenth centuries was characterized by the fact that it remained external to the natural sciences insofar as, on the one hand, the natural sciences themselves considered this discussion to be superfluous in view of the apparent success of their activities[2] and, on the other hand, those who were concerned with the fundamentals did not want to and could not in any way change or even improve the successful procedure of the natural sciences by giving it a foundation.[3]

Under these circumstances, the discussion of basic principles was suspected of stemming from philosophy's grievance that sciences formerly belonging to its disciplines, now appearing independent, could get by without its imprimatur; indeed, under these circumstances, the discussion of basic principles generally ran the risk of being discredited, of either imputing extraneous foundations to the disciplines to which it was unable to contribute and in which it was not even properly versed, or of extrapolating generalities from the procedures of these disciplines which merely amounted to a senseless duplication.[4]

The methodological uncertainties that arose for the natural sciences even in overcoming the classical physical world view created a change in this situation, first of all, so that the natural scientists themselves sought theoretical orientation.[5] However, in order to prevent possible illusions: these uncertainties did not irritate the practical course of the natural sciences to any great extent. On the other hand, what philosophy was able to offer the natural scientists in the way of theoretical orientation was not conducive to the natural scientists hoping that the solution of basic theoretical problems would solve the difficulties encountered in their disciplines.

If, nevertheless, in our century the foundations and presuppositions of the natural sciences are debated in a different way than in the eighteenth and nineteenth centuries, it seems to me that the main reason for this is that in our century the successful approach of the natural sciences themselves has become questionable,[6] first of all in

---

[2] The attitude of the chemist J. Liebig can be regarded as a classic example of this. Even if other natural scientists, such as Faraday, had hopes of a philosophical discussion of fundamentals, Liebig's attitude is, in my opinion, the fundamental tendency.

[3] This certainly does not apply to Goethe and Hegel; but in their position on natural science, these two are more likely to be the exception that confirms the rule and—as far as Hegel is concerned—have had to pay for it with all the contempt of natural science. In the nineteenth century, it is not only the so-called Neo-Kantians who confirm our observation, but also the heirs of Hegel.

[4] In my opinion, the basic approach of the so-called Vienna Circle or Reichenbach can also be understood as a reaction to this discredit.

[5] The school of Niels Bohr is an example of this, but in particular the theoretical questions urging philosophical discussion, which are grouped around the so-called Heisenberg indeterminacy relations.

[6] This scepticism has found expression in the scientific self-image in Max Weber's studies on Western rationality, in Freud's "Civilization and its Discontents" (1929, in Freud 1974), in E. Husserl's "The Crisis of Science as an Expression of the Radical Life Crisis of European Humanity" (in Husserl 1954/1970) or in J.D. Bernal's "The Social Function of Science" (Bernal 1939). If the natural sciences are only implicitly subject to this questioning, what is remarkable

an extra-scientific sense, which admittedly has intra-scientific consequences. It is sufficient for our understanding to visualize this extra-scientific questioning by referring to the atrocities of two world wars, by naming Auschwitz, Hiroshima or Vietnam. The infernal character of these catastrophes is not conceivable without the successes of the natural sciences and thus leads to the question whether there are reasons within these sciences themselves that make them jointly responsible for their contribution to those catastrophes.

In my opinion, this question cannot be ignored by the philosophical discussion of the foundations of the natural sciences; on the contrary, this question makes demands on these foundation in which, if they face up to them, they regain their justification and which can thus help them out of the bad reputation that has overshadowed their efforts in the past. In the following, I would like to put up for debate some considerations that seem to me to be important for the philosophical engagement with the foundations of the natural sciences with regard to these demands.

First of all, we should specify where a specifically philosophical contribution is required in this context. It is the use of scientific knowledge for unjustifiable purposes. The social sciences are the first to rise to the challenge of uncovering the causes of catastrophes and pointing out ways to create conditions under which such unjustifiable uses of scientific knowledge would be ruled out. As far as the natural scientists themselves are concerned, science is called upon to seek out the causes in the socialization of scientists, in the institutionalization of scientific work, etc., which must be held responsible for the compliance of scientists with such antihuman ends.

Philosophy, on the other hand, seems to be challenged to answer the question of whether this compliance also finds expression in the theoretical foundations of the natural sciences, such that this compliance is produced unnoticed in the execution of those sciences, as if it belonged to the matter itself. Philosophers, as is well known, do not agree on this question. Leaving aside the differences in detail and different camps as it were, two positions can perhaps be distinguished. First, there is the view that the findings of the empirical sciences can in principle only provide a means for purposes external to science as a science—anything else is a threat to their scientific character.[7] This view is opposed by those who either criticize the purported purpose-neutrality of the empirical sciences as illusory, or believe that said purpose-neutrality is anchored in its theoretical foundations.[8] Since only the second view claims to contribute something to the basic discussion in the sense defined here, I will deal with this one, whereby the first view is implicitly dealt with at the same time.

---

compared to the nineteenth century is that they are no longer excluded from critical scrutiny as something unassailable.

[7] With regard to the prevailing philosophical directions, it is much easier to cite so-called logical positivism and so-called critical rationalism as exemplary representatives of this position.

[8] For the German context, the so-called Frankfurt School should be mentioned here. In another way, however, it is also the so-called Erlangen School.

I do not want to begin by addressing the criticism of the alleged freedom of purpose, which centers on the argument that scientists cannot, whether by an act of will or by means of specific institutional provisions, escape the fact that they are conditioned by the historical-social reality to which they belong, but that, on the contrary, the attempt to remain indeterminate vis-à-vis this reality is an expression and affirmation of a certain role in this historical-social reality.[9] As a starting point, this line of argument has the disadvantage that it can easily appear merely as an external form of sociological rhetoric. For our question with regard to philosophy, therefore, another line of argument is more suitable as a starting point, based on epistemological considerations, which tries to show that cognition is unthinkable without an interest going beyond pure curiosity, mere thirst for knowledge. In a manner familiar to philosophy, this argument takes aim at the foundations of the natural sciences by claiming to show immanently, i.e. by analyzing the implications of every act of cognition, that in it, references to the object, which cannot be reduced to a pure interest in cognition, are constitutive, and thus belong to the foundations of cognition and cannot be overlooked as such constituents without consequences.

The main idea behind this argument[10] is that the subjectivity of all cognition should not be understood as something merely formal, i.e. not as something that only expresses that cognition is the cognition of a subject, while the content of cognition is determined solely by the object. Rather, cognition is to be understood as an activity of the subject, so that the content of cognition is determined at least by the object *and* the subject. A reflection on the foundations of a theory, so the constitutive subjectivity argument goes—presented as briefly as possible—which thinks of the contents of cognition as determined by the object alone, subordinates the subjective contribution to cognition to the object and thus withdraws it from all further reflection.

Now it is obvious that this idea calls into question the possibility of scientific knowledge if this subjective contribution is understood as something arbitrary and not something that, although subjective, at the same time has the character of necessity, even if only that of a conventional necessity. For the fact that cognition is subjective, when we do not mean this only formally, does nothing to call into question its bindingness, its objectivity and its form of generality. On the contrary, the core principle of all transcendentalist approaches is that the objectivity and general character of cognition are precisely based on the latter's subjectivity, whereby this subjectivity that guarantees objectivity can be understood as the form of apperception of each subject or as intersubjectivity. It is not possible to refer to empirical subjects, not only because otherwise a theoretical circle would be unavoidable, but also because of less strict considerations: Every subject necessarily has a particular point

[9] Cf. Lefèvre (1971).

[10] In the following, I do not refer to the position of a particular scientist or to that of a particular philosophical school. I have in mind here general transcendental philosophical trains of thought that interest me without further specification in their fundamental position on the problem of the foundation of science, especially since they may be illuminating for scientific positions that would never understand themselves as transcendentalist due to any of the specialities excluded here.

of view, which he considers to be true. The subject's interests, which differ from those of other subjects, are also decisive for this. If it is no longer possible to invoke reason as a dimension of subjectivity that resolves the contradictions between subjects, as was the case at the time of the bourgeois revolution, then a dimension of subjectivity that is transverse to the divergences of interests and thus cannot be shaken by them is required to guarantee the objectivity of cognition for divergent interests. This commonality of divergent interests can ultimately only be conceived in the form of a common means for different interests, so that the dimension of subjectivity that guarantees the objectivity of cognition must be called transcendental instrumental rationality. This transcendental instrumental rationality could be thought of as a necessary subjective (intersubjective) form of implementation of the constitution of objects through ends, a form of implementation that is not determined by those ends.

This solution is certainly not very satisfactory if we recall that the contribution of philosophy to the fundamental discussion of the natural sciences consists in questioning the purpose-free self-understanding of the natural sciences. For it cannot be denied that instrumental rationality, as a result of the transcendental approach, reproduces this purpose-neutral self-understanding, in that by means of it, science, which sees itself as purpose-neutral, is merely determined as science for every purpose.

This result discredits the argument that puts forward the subjective character of cognition also in the specific respect that it is not possible to see what sense there is in ascribing the predicate subjective to instrumental rationality. For what difference does it make whether I allow that necessary form of implementation of object-constitution to be grounded in the structure of the subject rather than in that of the object? Doesn't this transcendentalism amount to a determinism of cognition that cuts off the reflection of cognition much more fundamentally than naive realism, for whom the content of cognition is determined solely by the object? The danger of dogmatism that appears in transcendentalism is, as is well known, not merely a possibility of thought; I am reminded, for example, of Einstein's disgraceful codification of the concept of space, and time in classical physics in Kant's transcendental philosophy.

But let us go back to the intention of the constitutive subjectivity argument. By showing the subjective contribution to the cognition of reflection, it promised to regain something that would otherwise be subtracted from it, one would think that it was not a matter of establishing cognition through reflection as what it already was without reflection. The idea that not only the object determines the subject, but also the subject determines the object, nevertheless aroused the hope that the subject would secure a dimension of freedom in this line of thought.

The foundation of objective knowledge in subjectivity must be seen as a consistent form of assuring freedom at which this line of thought aims. The subject as constitutive of objectivity would at the same time be essentially free, because it would have understood itself as universally determining and only determined by itself in this foundation. This constitutive subject, however, would have to be conceived as distinct from empirical subjects for the reasons mentioned earlier. Apart from the difficulties associated with this, however, this constitutional-theoretical aspect of the argument contains a consequence that calls the whole argument into question as a critique of the purpose-free self-understanding of scientists that does not reflect the presuppositions of science. For this constitutional-theoretical aspect of the argument is by no means only directed against a passive-objectivist conception of knowledge, in which a certain justification could be conceded to it; it is equally, and in the meantime probably primarily, directed against the view that the practical, material relationship between man and nature is constitutive for knowledge, and that it cannot be a matter of deriving this relationship from theory, but only of deriving theory from this material relationship.

This constitution-theoretical exclusion of the practical, material relation between man and nature and thus of its respective historico-social form as the real constitutive of the theoretical relation to nature seems to me to be only the philosophical side show to the natural sciences which do not reflect their presuppositions and understand themselves as disinterested and purpose-neutral.

If this is correct, it would by no means be a negative result. Rather, it seems to me to indicate in a more specific way how philosophy can contribute to the fundamental discussion of the natural sciences. Generally speaking, one of philosophy's tasks is to show the natural sciences the historical-social form of the material relationship between man and nature as reflected in their theoretical foundations. To this end, I would like to present a few—admittedly only preliminary—considerations.

I want to start from a point of view that is also found in certain forms of constitutional-theoretical argumentation, namely the indication that natural science does not refer to nature as such, but to appropriated nature, and that it is only able to analyze nature that has not yet been appropriated with concepts that stem from the already completed appropriation of nature.[11]

The appropriation of nature can certainly only be thought of as a transformation of nature—the most primitive appropriation, eating, demonstrates this insight. But the transformation of nature through its appropriation is something other than constitution in the traditional philosophical sense. On the one hand, transformation is not constitution insofar as what is transformed was previously something definite; on the other hand, transformation is not the other of nature, but happens naturally itself, whether it is transformation of food through teeth, saliva and stomach secretions or transformation through hands, tools or machines. However, the transformational appropriation of nature through labor differs from all other natural processes of change in that it involves a conception of the desired result of change. This conception is thus constitutive for the result of the transformation; on the other hand, it will only have a result at all if it is an expression of the respective historical state of mastery of natural formative potentials. This conception does not constitute these natural forming powers, nor the state of their application by humans, but rather presupposes it.

[11] See e.g. Kambartel and Mittelstraß, eds. (1973).

At this point, the natural scientists, rightly in my opinion, do not allow themselves to be talked out of their realism, but stick to Francis Bacon's motto that we can only control nature by obeying it. It is, of course, the task of philosophy to show natural scientists, right down to their most elementary concepts, that their findings do not simply reflect untouched nature, but nature as a power of formation on the basis of an appropriation that forms it. The historical status of this formative appropriation of scientific conceptualization is by no means external, as can be seen, for example, in the concept of force in classical physics, which is inconceivable at a stage of development at which human and animal bodily forces are the essential forces of mechanical formation.

In view of this historical perspective, perhaps also the concept of constitution receives a relative justification. Subjective idealism, to which it belongs, probably did not develop by chance at a time when the relationship between man and nature underwent a qualitative change. On the basis of agricultural production, which was dominant until the seventeenth/eighteenth century, human labor appeared primarily as a supporting element in the execution of natural production processes that were vital for humans (plant growth, animal reproduction); labor appeared as the promotion of natural production processes that took place naturally. In non-agrarian production, on the other hand, which displaced agriculture from the first place around the time of the seventeenth/eighteenth century, work—in principle even in the most primitive hand axe production—is the setting in motion of natural formation processes that do not take place naturally, but to which work must force existing natural things, arranging them as a result. Here, the image of a nature that is brought out of its inherent course and constituted through practice as a nature that forms itself for human beings is already laid out *in nuce*, an image that realistically reflects the actual developmental tendency since the eighteenth century.

The creation of this second nature presupposes a degree of mastery of nature objectified in means of labor, the application of which demands certain forms of labor. It is no coincidence that the transition to industrial production is a process of the dissolution of individual family production and the formation of social labor in manufactories and factories, the size of which is admittedly dwarfed by today's corporations and trusts, in which sometimes a million people work together under one roof. This enormous revolution in the form of labor in the last 200 years is just as much a prerequisite and condition of a practical relationship to nature as the development of the means of labor and the corresponding qualification of workers, in which nature comes into being as a producer for people; it is thus also a prerequisite and condition of the theoretical relationship to nature and by no means a circumstance external to it. This is not only evident in the fact that the figure of the individual researcher has largely become an anachronism; it is equally evident that industry based on social labor has become the direct precondition of natural research.

The social form of labor is thus—philosophically speaking—the condition for the possibility of the mastery of nature and thus also of natural science. At the same time, it shows us that a certain state of mastery of nature requires certain social forms of labor. The forms of labor are by no means subject to our arbitrariness. The present state of the mastery of nature makes planned social work necessary, not only on the level of a large-scale enterprise, but in the dimension of society itself.

It is against the insight into this very necessity that, in my opinion, the point of the purpose-neutral conception of science is directed today. The conflict of interests that this view wants to keep at arm's length from science can no longer be understood as an irrational chaos of individual interests and value orientations; in my opinion, for about a hundred years it has been essentially about the alternative: to use planned social labor or not.[12] For the natural sciences, neutrality in this question therefore means remaining neutral with regard to their conditions of existence and development.

The contradiction between the purpose-neutral self-understanding of the natural sciences and their conditions also reveals itself as a contradiction in their theoretical foundations. On the one hand, knowledge of a thing is considered true if it can systematically guide an action to produce the thing; it has no other criterion of truth, as experiment or deduction shows. In this way it reveals that what it recognizes is actually or potentially appropriated nature; that the natural forces it investigates begin to pass over into human forces precisely as forces under investigation. And yet, on the other hand, it is taken for granted that it has to do with nature as such, that its own character as a productive force is something external to it, something that belongs to it only possibly and accidentally.

These considerations may seem to a natural scientist as an idle and just therefore typically question of philosophical nomenclature. For whether I understand a law of nature as a human potency of production or not, its specific content seems to be unaffected by it—iron is obtained by reduction from ore anyway. At most, one will admit that such considerations can contribute to stimulate the reflection on the social conditions of the natural sciences.

It is questionable, however, whether the view that such reflection ultimately remains external to the natural sciences can be described as appropriate to the present situation in which scientific planning has become a necessity.[13] In this planning, it emerges not as a philosophical but as a practical principle that natural science should not be detached from the laws of nature in its research.

Nature that has already been appropriated is not, as in earlier times, only a prerequisite for further appropriation of nature. It not only defines the possibilities of this further appropriation, but also confronts it with an alternative. This alternative concerns natural science as such from the point of view of the enterprise, and can well show that it is not without consequences if it is abstracted from the fact that the appropriated potencies of nature are potencies of human beings that can only be realized in a certain social form, namely in the form of planned social labor. For as human potentials not realized in planned social work, the appropriated forces of nature are uncontrolled like non-appropriated forces of nature and destructive like

[12] Cf. Max Horkheimer's essay "Traditional and Critical Theory" (Horkheimer 1937, translation Horkheimer 1992), reprinted in Horkheimer 1988.

[13] Cf. P. Weingart's introduction in Weingart ed. (1972).

them. The basic assumption of the natural sciences, that they only deal with nature as such, unintentionally gives expression to this and at the same time obscures the insight that it is due to the unimplemented social conditions of realization if Second Nature, which they decisively helped to create, does not differ from First Nature in the all-important point, namely to be uncontrollable and catastrophe-prone for human beings. This concept of nature obstructs the realization of the mastery of nature, i.e. of the program of natural sciences, by cutting off the scientific penetration of the social side of this program.

Immanently formulated: the claim contained in the purpose-neutral self-understanding of the natural sciences, to provide people with means, is thus shown to be excluded by the exclusions of this self-understanding itself. The natural sciences thus define themselves as ends in themselves, professing the motto "Long live science, may the world perish from it!"—a motto which in our century, thanks to the achievements of the natural sciences, is to be taken in the literal sense. This cannot be remedied with an oath of Hippocrates and not with the appeal to enter into a discussion of purposes free of domination; for in this way, as I have tried to show, it would again be excluded that the natural sciences and their findings are indissolubly interwoven with the productive forces and themselves belong to them, which at their present level of development demand a handling corresponding to their social character, i.e. socially planned, if they are not to turn into forces of destruction. This seems to me to be the core of the problem, and in the misrepresentation of this core I see the ideological character of the concept of nature of the natural sciences.

The exposure of this ideological character, its causes and its consequences seems to me to be a contribution of philosophy to the basic discussion of the natural sciences, which could do justice to the extra-scientific problem mentioned at the beginning, which the people of our century have with the natural sciences.

Appendix B

Between Two Natures
Wolfgang Lefèvre
Zwischen zwei Naturen. In: H. Vierck (Hg.): *Der Comenius-Garten–Eine Leseprobe aus dem Buch der Natur*, 49–53. Berlin (Edition Hentrich).

The book of nature is written in the language of mathematics. With this oft-quoted dictum, Galileo in his *Saggiatore* identified a topos that can be traced back to the origins of Christianity: the idea that God not only revealed himself in the Holy Scriptures, but also in his creation, the "book of nature." But do these books teach the same thing, and which takes precedence when they seem to contradict each other? The idea of the two books, which expresses the conviction of Augustine and other theologians that faith and knowledge could not contradict one another in the last instance, in fact staged a conflict between these two spheres as well as laying the groundwork for the theological regulation of science. It may therefore seem remarkable that Galileo of all people—he was after all the founder of modern science—should appeal to this double-edged metaphor. Indeed, he was playing with fire—Bruno was burned for such heresy in 1600—by insisting that only the mathematically educated were equipped to interpret the book of nature. Why, we might ask, did Galileo not take the path offered by Averroes, whose doctrine of double truth proposed that knowledge of nature was of a different order of truth from the knowledge of theology? Perhaps for Galileo it was simply clear that natural science, unlike theology, is not concerned with the meaning and purpose of nature as creation, but only with the way nature actually works. Or perhaps his reference to the book of nature indicates a conception of nature and natural science that is foreign to us today.

In his *Physica*, Galileo's contemporary Comenius presents matter, spirit, and light as the fundamental principles whose interaction produces and sustains all natural phenomena, which return to these elements when their existence ends. At first such a proposition may seem strange from a contemporary perspective, but on closer inspection its meaning becomes clearer. When Comenius speaks of the violence of fire (light), which prepares inert matter to be shaped by spirit, it is helpful to remember the Paracelsian chemistry of his time, which considered fire the universal means by which substances were decomposed, a means whose mode of action was imagined to be quite mechanical in the seventeenth century. Light was also thought to fall on the spirit from the volatile distillation products known as spiritus, with whose invigorating and healing effects Comenius the physician-chemist was concerned. Such additional knowledge of the history of science equips us to approach Comenius' explanations of individual natural phenomena—for example, states of aggregation, changes in density, etc.—with more certainty, to find out how those fundamental principles are conceived. From this perspective, the 'wrong' explanations are especially interesting, because reconstructing why they were right for Comenius allows us to access his theory of nature.

It is often assumed that Comenius tried in his *Physica* to do the same as today's natural scientists under different conditions and circumstances and with other—from today's point of view insufficient or even failed—means. But what if this assumption is mistaken? What does one then do with remarks like these: "These three (matter, spirit and light) are therefore the truest principles of all in our world. Matter is the true mother of all; spirit the true father of all; and light, however, as it were the common love of both, the true bond between things and the heritage of love for the descendants of all creatures; always active, maintaining all things and itself in perfect existence, and propagating them further. This is the basis of the eternal harmony in the preservation of things." Is this to be taken merely as a pious or effusive appendage to his *Physica*, which does not touch its scientific core and can therefore be omitted without changing anything essential in Comenius' work as a document of the study of nature in a past period? Or, on the contrary, is it only on the basis of such explanations that we can understand what the study of nature of men like Comenius was about, and what distinguishes it not only in its appearance but in its very core from today's natural scientific research?

To read the book of nature, that is, to study it as a revelation of God's thoughts and purposes, as an "admirable image of the eternal mystery" (Comenius), as a "viva Dei statua" (Campanella) or, in secularized terms, as an absolute idea (Hegel), means—to put it as moderately as possible—to seek to understand nature as a meaningful whole. From this perspective, to study nature—with the subtitle of Bacon's *Novum Organum*, which Diderot took up again—was to *interpret* nature; to read the book of nature was an act of exegesis, an effort to understand nature's meaning. Those aspects of such an endeavor that look like today's natural scientific research—the attempt to explain the occurrence of certain phenomena and the mode of action of certain processes in nature—were always also conducted in this light. It was not because the geocentric astronomy of Ptolemy had failed to reconstruct certain celestial phenomena or was inadequate in its prognostic potency that Copernicus and later Kepler designed a heliocentric alternative, but because Ptolemy's system, with its eccentricities and epicycles, did not produce a well-ordered, meaningful picture of the natural world.

We readily grant such grand narratives a positive role in the formation of the modern natural sciences because they valorized nature; they envisaged nature as created by God in such a way that, once created, it reproduces itself and can thus be understood only on the basis of its own laws. Nevertheless, such general philosophical orientations remain for us religious remnants or traces of the "metaphysical stage of development" of human cognition (Comte), which natural science had to strip off in order to become truly "positive science." But what were these orientations based on, and what made it possible to transcend them?

In our everyday lives, people have a practical relation to nature prior to any reflexive relation, and our practical relations to nature change historically. Stone Age hunter-gatherers related differently to nature than did the farmers and cattle breeders of ancient or medieval societies, and these in turn differently than craftsmen or later inhabitants of industrial societies. Hunter-gatherers appropriate what nature presents them with in the way of food; farmers encourage the natural processes that produce food; the artisanal producer uses nature as a storehouse of working materials which he uses to produce his sustenance and livelihood; and the industrial producer arranges nature into a new world of aggregates that produce for human purposes. People treat nature as something different in each of these different practical forms of life, whether they realize it or not. When they reflect on nature, they do not think about nature as such, but about nature as they actually encounter it in their practical form of life.

From the Neolithic Revolution to the Industrial Revolution, agricultural production was the decisive basis of life in human societies; the cultivating life practice, as I will call it, dominated all others. I call this life practice "cultivating" because in animal and plant breeding the working man promotes and cultivates certain natural production processes, namely the natural reproduction of certain animal and plant species. It is not the farmer that produces his food, if by production we mean manufacture. Rather, natural beings produce food for the farmer out of their own natural life activity, which the farmer merely supports with his activity. Even if he intervenes massively in nature, cultivates areas of land, clears forests, drains swamps, transfers plants from one climate to another, etc., the farmer thereby only prepares the ground for the natural producers of his food. On this model, the working man is in the last instance only an assistant of nature; he makes himself and his labor into a moment in the natural production process. In this life practice, nature is a context of life-giving processes into which people fit and which they must preserve insofar as they are able to, however precarious the balance may be between stewardship and potentially or actually destructive intervention.

Where the cultivating life practice prevails, nature with all its horrors appears to people as a life-giving cosmic process, and it is therefore quite realistic when people in ritual, mythological or later philosophical representations of their way of life depicted, celebrated or conjured up nature as a purposeful whole, a fateful connection of processes the meaning of which must be grasped in order to lead a successful life by participating in nature accordingly. On this view of nature, it could only have seemed presumptuous, even downright wrong, not to obey nature for the sake of a life in accordance with its workings, but to try to control it, to use Francis Bacon's classic formulation.

It is a different nature that Bacon has in mind, the one that artisanal—and later industrial—producers deal with. In this productive life practice, human work does not nestle within nature conceived as a given whole of life-giving processes, but takes given nature as raw material that can be used to establish new contexts of action for human purposes. I call this life practice "productive" because in it people themselves are producers of their food. Of course, the artisan producer also exploits natural production potentials, even if it is only the effect of natural law that distinguishes between the hardness of tool and workpiece. But where the artisan's tools appear only as a further extension of his bodily organs, so too his activity is the sole moving principle of production, which must conquer an inert, merely passive or even resistant nature. And even when it is unmistakable that the artisan cunningly coerces nature into serving his purposes, for instance by means of so-called simple machines (our words "machine" and "mechanics" are derived from the Greek *mechane*, which means not only tool, invention, etc., but also cunning), the processes thereby put into action appear not as natural, but as artificial.

Until the Industrial Revolution, this productive life practice was embedded within the hegemonic cultivating one and gained a certain independence from it only temporarily in urban societies. Under such exceptional conditions, however, it left its mark on the image of nature encountered in the cultivating life practice: the life-giving cosmic process was increasingly seen as a mechanical one; the world could increasingly be thought of as a machine and creative nature as a studio. We find such a two-sided image of nature in the ancient city-states and especially in the Western and Central European societies at the time of Comenius, in whom we read: "For realization to take place (*ad actionem*), three things are necessarily required: something to be realized, or an object; an actor, or subject; an instrument or tool. The physical world, because it had to be full of effects and states, of procreations and changes, necessarily required those three things. But behold, they are there! The constant and general object is the substance poured out over all; the constant and general actor is the spirit poured into all; the general instrument, constantly moving and active, is the light penetrating all, changing all."

Though one can still read the trace of the image of nature that dominated the cultivating life practice in these remarks, despite all the spirit of craftsmanship, the end of this image's dominance is announced at around the same time in Bacon's dictum. The decisive factor was not least the fact that, even if production nevertheless remained essentially manual until the nineteenth century, in Bacon's time machines were beginning to be used in industrial production to an extent unknown to antiquity. Whereas mechanics had until then been regarded as a science of art rather than nature—mechanics was not a branch of, but something fundamentally different from physics—a distinction now began to be made within mechanics between the "forces of the hand" and the "forces of nature" (Newton). Although natural forces were still conceived of as natural, they could increasingly be accessed via human inventions, such that forces, which could be used and studied by machines, began to shape human conceptions of what nature is: a realm of productive potencies by means of which man makes himself the master of nature.

The industrial mode of production pushed these developments to their conclusion. By taking nature's productive potentials into service, humans arranged nature into a Second Nature, into a world of aggregates produced and reproduced for human purposes. In this mode of production, nature has no meaning or purpose for natural scientific research to decipher. From the perspective of this life practice, the study of nature serves only the appropriation for human purposes of nature's productive potential; the sole purpose of given nature is to be transformed by human beings.

It would be, to put it mildly, a fruitless undertaking to present a garden, itself an arrangement of nature, as an instance of that meaningful First Nature that dominated the worldview of the cultivating practice of life. Yet a garden can draw attention to the fact that the Second Nature we have created is far from making human sense.

Appendix C

Knowledge Systems in Historical Change
Peter Damerow
Wolfgang Lefèvre
Wissenssysteme im geschichtlichen Wandel. In: F. Klix und H. Spada (Hg.): *Enzyklopädie der Psychologie - Themenbereich C: Theorie und Forschung* - Serie II: Kognition, Band G : Wissenspsychologie, 77–113. Göttingen (Hogrefe).

Individual and Social Knowledge

The psychology of knowledge is primarily concerned with the *individual's knowledge*, studying above all individual mental processing (acquisition, storage in memory, retrieval, exchange, application, etc.) (Mandl and Spada 1988). The question as to how knowledge systems change historically differs fundamentally: it is not concerned with individual knowledge but with *social knowledge*, i.e. it refers to knowledge only insofar as it is shared intersubjectively by numerous individuals and passed on historically with a certain continuity. Investigating this social knowledge, as well as its contents, structures, and the mechanisms and forms of its transmission (school knowledge) is therefore not originally a psychological but a sociological and historical task, which cannot be accomplished using the methods of empirical psychology (Berger and Luckmann 1966). Psychology is limited in principle with regard to social knowledge, because the functions and structures of knowledge that is intersubjectively shared and transmitted are determined—at least in part—by mechanisms of interaction between individuals and their external conditions, and not only by the human psyche.

The historical development of knowledge systems is of constitutive importance for psychology, however, because connections exist between the mental forms of knowledge processing and the content of knowledge (Weinert and Waldmann 1988). Not only the content of knowledge, but also the structures that organize, fix, and store knowledge are subject to considerable historical change (Wozniak and Fischer 1993). It cannot be ruled out, therefore, that mental processes and cognitive structures, i.e. the original objects of psychology, are also directly affected by historical developments.

This is why, even after Wilhelm Wundt's monumental *Völkerpsychologie* (1900–1920, see especially vol. 10: Culture and History), which marks the transition from humanistic to empirical psychology, there have been repeated attempts to address the acquisition, mental representation, and application of knowledge in past historical periods (Stern 1920). A classic study in this respect is Wertheimer's investigation of productive thinking (Wertheimer 1945/1957), which contains an analysis of Galileo's discovery of the principle of inertia and a psychological reconstruction of the heuristic considerations that led Einstein to formulate the special theory of relativity (Wertheimer 1920, Wertheimer 1945/1957). Yet these case studies were intended more as illustrations of Wertheimer's theory and, despite important insights, generally hardly stand up to an examination of the historical facts in detail (Damerow et al. 1992, on Galileo; Miller 1984; Renn 1993, on Einstein). This is particularly true of works in the Herbartian tradition, in which the recapitulation thesis, "ontogenesis recapitulates phylogenesis," for a while became a widespread model for interpreting the relationship between social and individual development. The historical examples with which Piaget sought to underpin his theory of the psychogenesis of basic categories of scientific thought are similarly problematic, especially his interpretation of the history of the concept of impetus (Piaget 1978; Wolff 1978), and the examples in the comprehensive study on the history of science published together with Garcia only after Piaget's death (Damerow 1993b; Piaget and Garcia 1989). Another broad study containing psychological interpretations of historical developments is Klix's (1993) investigation of the developmental phase of human cognitive performance from the human-animal transitional phase to antiquity, as well as his investigation of the nature of the mind (1992). Psychological studies also occasionally address historical developments concerning specific objects, especially the concept of number (Brainerd 1979; Dehaene 1992).

Occasionally, psychological studies use historical examples to illustrate cognitive structures (mental models, etc.) that still exist today (Gentner and Stevens 1983; Giere 1988, 1991, 1992), for example those concerning the concept of force (Clement 1983; McCloskey 1983), the concept of movement (McCloskey and Kargon 1988) or the concept of warmth (Wiser 1988; Wiser and Carey 1983) (cf. also Vosniadou and Brewer 1987, 1992). Finally, there are psychological studies on the cognitive processes of individual researchers, such as Gruber's study of the development of Charles Darwin's ideas (1981).

Mental (Internal) and Cultural (External) Representation

If the historically changing cultural environment of individual human development determines not only the contents of knowledge, but also the mental functions and structures of knowledge systems as they are individually constructed (i.e. if social knowledge significantly influences the construction of the respective individual knowledge), then this presupposes that social knowledge is not only presented and transmitted *mentally* in the individual, but also *culturally* through historically changing artefacts, especially in the material means of communication and intellectual activity (tools, cult objects, pictures, signs, language, writing, etc.). The process of representing and transmitting the knowledge manifested externally in artefacts in turn has psychological preconditions, but it is by its nature a phenomenon that must be understood historically and sociologically.

This results in the fundamental problem that psychology can only define culture as an accidental boundary condition of individual development and include it in the study of cognition (Bruner et al. 1966; Poortinga 1992). Psychology primarily studies *mental* representation (Engelkamp and Pechmann 1993; Mandl and Spada 1988; Opwis 1992; Stevenson 1993) in its various forms (figural, imaginative, sequential, in semantic networks, in logically structured systems, as "frames," as a production system, etc.). These forms of mental representation have external counterparts in *cultural* representation in the form of real objects and actions (images, processes, forms of interaction, forms of language, written representations, etc.) (Aebli 1980/1981; Damerow 1993a). The intersubjective nature and historical continuity of the functions and structures of knowledge systems beyond the psyche of the individual subject are guaranteed by the existence of these objects and actions, which is independent of mental representation, and by the social tradition of their meanings.

Psychologically, external representations have the same functions as internal representations and are based on the same mental abilities to use symbols (Werner and Kaplan 1963). From the perspective of psychology, external representations therefore appear as externalizations of human knowledge by means of culturally predetermined media of representation, whereby enactive, iconic and symbolic representations can be distinguished according to the type of media (Bruner et al. 1966). From a historical-sociological perspective, on the other hand, mental representations appear as internalizations of the objects and actions in which knowledge is externally represented.

Bruner attempted to extract explanations for culturally specific cognitive operations from this assumption (Bruner et al. 1966, Chap. 14). Further conclusions have been drawn by the cultural-historical school, according to whose theory all fundamental psychological processes (concept formation, deductivity, etc.) have a historical character and are shaped by the respective concrete social-historical conditions (Cole and Scribner 1974; Leontjew 1973; Luria 1971, 1976; Wapner and Kaplan 1983; Wertsch 1985; Vygotsky 1934/1962). From this theoretical perspective, tool use and language are constitutive for human development and form the prerequisites for the emergence of *culture*.

In any case, historical changes that the functions and structures of knowledge systems undergo can be explained by analyzing the way in which the specific means of communication are used in the various historical epochs, as well as the forms of intellectual labor that take place within them, both of which—the means of communication and the products of intellectual labor—embody the culture of these epochs. In this way one can explain historical changes that are not caused by the nature of individual cognition and can therefore be confirmed—but not theoretically reconstructed—through comparative historical and cultural analyses using the methods of cognitive psychology. On the basis of the cultural-historical context, psychological functions and structures of knowledge systems can primarily be explained because the use of such means makes historically changing mental demands, and requires the transmission of specific knowledge for their adequate use. In addition, the form of the mental representation of knowledge itself (pictorial, propositional, rule-based, etc.) is structured by and acquires certain functions from these cultural representations.

In order to clarify the historical conditions under which the psychologically determinable structures and functions of different knowledge systems have emerged, the factors determining the cultural-historical context of knowledge in the various phases of the development of cognition must be examined via the findings of historical disciplines. This includes the objects, structures and functions of intersubjectively shared knowledge, its external representations, the social mechanisms of its transmission as well as its function in the respective system of social reproduction.

Knowledge Systems in Oral Cultures: Knowledge as a Functional Moment of Action

Insights into knowledge in oral cultures are made available above all by ethnology and anthropology. These disciplines acquire their knowledge not only by researching recent peoples with a corresponding form of civilization, but also—in connection with pre- and protohistoric archaeology and historiography—reconstructed from the legacies of past oral cultures or from historical reports about them. However, the problems associated with such reconstructions and with understanding foreign cultures make these research results inevitably fraught with a certain degree of uncertainty, especially when it comes to the intellectual activities of oral cultures. Nevertheless, some basic features of knowledge at this level can be regarded as certain.

First of all, with regard to the objects of social knowledge in oral cultures, it must be noted that an often astonishingly rich and detailed knowledge of natural facts is connected with the skills and technologies used to sustain life. This is particularly evident in the differentiated nomenclatures for animal and plant species found in such cultures. For example, comprehensive and highly differentiated taxonomic distinctions of animal and plant species have been documented for Stone Age cultures on New Guinea (Diamond 1966; Hiepko and Schiefenhövel 1987), which show extensive similarities to modern scientific taxonomy (Gould 1980). Another area of rich knowledge relates to the culturally differentiated, socially standardized behaviors associated with certain situations and social roles or among certain age groups. The partly counterintuitive, artificial gift-giving and exchange rituals, which were already reported in the classical ethnological literature (Malinowski 1966) and continue to form a central object of more recent ethnological studies (Bell-Krannhals 1990; Schiefenhövel and Bell-Krannhals 1986), offer an example of such socially standardized behavior, which can be seen as extreme from the perspective of the cultural dependency of social cognition. Both in the case of biological nomenclatures and in the case of gift and exchange rituals, the contents of knowledge cannot be traced back to simple, individual learning processes, but the acquisition of the stock of knowledge is linked to historically and culturally specific framework conditions of the individual's development. The same also applies to mythologies as another area of extensive knowledge in oral cultures.

At this level, knowledge is both communicated and transmitted from generation to generation in an exclusively interactive form. Spoken language is the most important means of communication, and the transmission of knowledge accordingly takes place to a large extent through oral instructions and narratives (Goody 1987). However, linguistic communication is accompanied by a sophisticated practice of knowledge acquisition through imitation in the context of shared activities, and participation in work processes and rituals—not a separate transmission system—is a key element of the transmission of knowledge (Alt 1956). The communication of knowledge is also often supported by symbols (masks, etc.), whose meaning is usually only determined holistically. Only in isolated cases are symbols already used with precise meanings, for example counting devices such as number word series or body-part counting systems (Saxe and Posner 1983), and graphic representations such as route sketches or geographical signposts (Dröber 1903/1964), an impressive example of which are the so-called toas of the Australian aborigines of the Lake Eyre basin.

Another characteristic of knowledge systems in oral cultures is the fact that knowledge is usually generally accessible. However, due to the close connection between knowledge acquisition and participation in certain activities, especially as a result of forms of division of labor, considerable differentiation can already occur with regard to knowledge and associated skills (approaches to expertise, use of the life experience of older people in councils of elders, etc.), whereby the fact that certain spheres of life are taboo for excluded members of the collective (e.g.

women's shelters tabooed for men) can even be connected with the establishment of secret knowledge.

Finally, specific forms of knowledge organization and integration are characteristic of knowledge systems of oral cultures. Since at this level of culture, knowledge, due to its close connection to concrete actions, refers in principle to real objects and facts instantiated in real actions, knowledge systems predominantly relate to specific domains, and knowledge is not subsumed in overarching abstract categories. This may even apply to mythological knowledge: it is doubtful whether the animistic or magical world interpretations of the so-called nature religions can be regarded as an overarching form of knowledge integration. It is more likely that they are culturally specific projections of motives and possibilities for action that dominate in the respective practice of life (such as the "eternal hunting grounds" of Native Americans).

In oral cultures, knowledge is thus predominantly a functional moment of action. In general, it is situationally bound knowledge that is communicated interactively. Admittedly, in spoken language knowledge is represented independently of the respective action context, but not in such a way that it transcends the situation. At most, the use of symbols points to an external representation of knowledge that transcends the situation, but the meaning of symbols is usually holistic, so that their cognitive function can hardly be compared to the symbol systems of later written culture.

The possibility of building up and passing on complex knowledge systems is thus very narrowly limited in oral cultures. This becomes clear in certain isolated, outstanding cognitive achievements, which also deserve interest because they represent transitions to the forms of knowledge of a higher stage of development. An exceptional example is the well-documented technique of navigation among the natives of the Caroline Islands in the western Pacific. The unusual navigational achievements of this technique would seem to be based on an astronomical and geographical knowledge that is comparable in structure to the basics of navigation in the Western European tradition. In fact, however, it is a borderline case of extensive use of the forms of organization of situation-specific knowledge characteristic of oral cultures, which remains bound to the specific geographic circumstances in which it is acquired and transmitted (Gladwin 1970; Hutchins 1983). These navigational achievements are also based on external conditions of learning that are extremely atypical for oral cultures, namely on a form of specialization independent of the social hierarchy, with pronounced selection criteria based on performance and years of practical training organized in a school-like manner.

Knowledge Systems in Early Written Cultures: The Symbolic Representation of Knowledge

The emergence of the early written cultures (Egypt and the Near East before Hellenism, ancient China, ancient American cultures, etc.) represents a significant turning point in the historical development of knowledge systems. Surviving written testimonies provide an insight into the body of knowledge of these cultures. The knowledge documented by such written testimonies, which is different from that of the preceding stage, has become the subject of research in the history of science. Such research, however, is not primarily oriented towards questions of psychological interest. For this reason, the psychology of knowledge must also draw on the results of the respective research in ancient history, especially the relevant philological disciplines.

The first thing one notices at this cultural level is the considerable expansion of systematically transmitted fields of knowledge. The extensive administrative knowledge, the development of complex metrological systems (weights and measures, calendars), the codification of legal knowledge (casuistic legal texts), and a form of proto-historiography (royal inscriptions and lists), which can be regarded as an early form of historical knowledge, are innovations. The development of areas of specialized knowledge is also new. For example, we have evidence of cross-cultural geographical knowledge; of casuistic medicine such as the ancient Egyptian medical papyri; of systematized magic such as the Mesopotamian omen and oracle texts; of astronomical knowledge, such as documents with astronomical events in the Mayan culture and especially in Babylonia that were recorded over many years and in part already processed mathematically; and last but not least, evidence of (pre-Greek) mathematical knowledge, for example, the mathematical cuneiform tablets, the ancient Egyptian mathematical papyri, the calendar calculations of the Mayans and the mathematical texts of ancient China.

On the other hand, the lack of overarching integration of such knowledge must be noted. In these cultures, mythologies, which are no longer only passed on in narratives but are available in written epics, theogonies, etc., bring almost all areas of life into a reflexive relationship. Yet it is not so much factual contexts that are reflected in mythology as political and state integrations, which are expressed as identifications of gods, as genealogical links between different religious traditions, or as transformations such as the patriarchalisation of the heavens of the gods.

All these fields of knowledge of the early written cultures are the knowledge of experts, i.e. elites, which the historically new division of labor between manual and mental work produced. Against the general background of the employment of foreign labor on a grand scale (enslavement of conquered peoples, etc.), a socially differentiated planning activity developed in the first cities that were emerging as the centers of early state formation. The experts of this centralized planning system (administrative and priestly officials, scribes etc.) organized the provision of materials required for agricultural and artisanal labor, as well as the distribution of this labor and its products. The social function of this differentiated intellectual work explains the peculiar canon of the aforementioned new types of knowledge. As a rule, this expert knowledge is not production knowledge (technical knowledge, knowledge about natural processes that can be used in production etc.), but primarily knowledge about the social organization of production (knowledge of domination).

The planning activity from which this novel knowledge emerged would not have been feasible without its external representation that transcended the situation. It is no coincidence that the emergence of precise forms of geometric representation (ancient Egyptian building drawings, Babylonian field plans, etc.) is an innovation of this period. It was the invention and development of writing, however, that played the key role in implementing this planning activity. Contrary to conventional opinion (Gelb 1958; Sethe 1939), writing was probably first developed in all early written cultures as an administrative instrument in the form of number sign systems and accounting scripts, or adapted from other cultures: see, for example, the pre-written administrative symbols of the Near East (Schmandt-Besserat 1992), the proto-cuneiform (Boehmer et al. 1991/1993) and proto-Elamic (Damerow and Englund 1989) administrative texts, the Minoan administrative texts (Chadwick 1958; Hiller and Panagl 1976), the Inca knot-cord technique (Ascher and Ascher 1972), etc. It seems that writing generally only became a means of representing spoken language later, when it developed into phonetic writing and acquired the additional functions of communication (letters) and the transmission of knowledge (beginning of the literary tradition).

At the same time, representing knowledge in permanent symbol systems allowed knowledge to be accumulated in new ways (Goody 1986) through the transmission of records (emergence of written archives). The most significant change in the transmission of knowledge that resulted from the use of writing as a means of representation was, however, the establishment of schools in which future administrators acquired the knowledge they needed separately from the execution of planning activities (Kriss-Rettenbeck and Liedtke 1986; von Hohenzollern and Liedtke 1988).

As far as the structure of this new type of expert knowledge is concerned, it should generally be emphasized that it was only indirectly related to real objects and facts, namely through symbolic actions (writing and reading, bookkeeping, etc.). The direct reference of this knowledge was the permanent symbol systems used, through which it was passed on. These, in turn, represented the processes of planning activity (speech acts, administrative acts, etc.) to which the new expert knowledge ultimately referred.

The most striking consequence of basing knowledge on symbolic actions was first of all a canonization of meanings, as is visible, for example, in the archaic lists of proto-cuneiform script (Nissen et al. 1993), and finally the emergence of abstract technical terms and languages whose meaning was not directly derived from real objects, but was first constituted by the structure of symbolic actions. This is how, for example, calendar terminologies such as those of the Maya (Thompson 1960), Mesopotamian administrative terms (Englund 1990; Nissen et al. 1991), and the terms of ancient Babylonian mathematics (Høyrup 1990) etc. came into being. Thus, the new expert knowledge of the early written cultures is related to the means of information storage and processing, which is also largely structured by these means of external representation.

The fact that knowledge was closely bound to certain symbolic forms of its representation in these systems also helped to establish the limits of such knowledge. These limits are expressed above all in the lack of a reflexive structuring of knowledge. The means of external knowledge representation generally seem to have been used directly, i.e. without any reflection on the symbolic actions performed. The practice of the experts largely determined the structure and function of this knowledge.

For this reason, research in the history of science has for a long time failed to recognize the character of this knowledge, which is quite distinct from practice, and the surviving evidence of this knowledge has been misinterpreted psychologically in certain ways. For example, compilations such as the collections of exercises in the ancient Egyptian mathematical papyri and in the ancient Babylonian mathematical cuneiform texts have often been interpreted as the expression of mere *recipe knowledge*. However, the various forms of pre-Greek mathematics are examples of a tendency towards an independent use of the symbol systems of differentiated intellectual work, i.e. a use independent of the execution of certain planning tasks. Institutionalized teaching in schools, which was relatively independent of practice, was probably responsible for the increasing independence of knowledge structures (Høyrup 1980). It can be seen as the beginning of attempts to explore the possibilities of using knowledge that the symbolic systems provided, which necessarily entailed a more reflexive use of the symbolic means of representation. This process of exploration can be called *scientific* knowledge production (Damerow and Lefèvre 1981, 223–233).

Knowledge Systems in Classical Antiquity and the Middle Ages: The Independence of Knowledge Production

The reflexive use of symbolic means of representation marks a new stage in the historical development of knowledge systems, expressed in the literature of the cultures of classical antiquity (Greece, Hellenism, Rome) as well as the Arab and Christian Middle Ages.

Once again, the transition to this stage of development of knowledge systems is marked by an expansion of the systematically transmitted fields of knowledge. Descriptive natural history, technology, history, regional geography, literature and religions emerged as differentiated objects of knowledge that were now fixed in their own literary forms. Areas of knowledge that developed during the previous stage, such as law and medicine, were also systematized during this period. The most striking, novel characteristic of this stage in the development of knowledge systems is, however, that the latter were now available for the first time in the form of theories. This applies, for example, to the deductive mathematics of the Euclidean tradition, to mathematical geography, to astronomy and to music theory, as well as to meta-theoretical knowledge systems, such as logic and linguistic theory. Finally, philosophy, which emerged and developed at this stage, and which is closely connected with the development of universal religious world views with a theological foundation—as expressed in Aristotle's "unmoved mover" or in the theologized Judeo-Christian God, etc.—represents, if not the first, then at least a completely new form of integrating the stock of knowledge.

This new kind of knowledge is also expert knowledge, but expert knowledge that no longer has a direct function in the exercise of social rule. It is an independent knowledge that is detached from work and planning processes and, in view of these processes, is also largely useless, which is why its bearers used the term "liberal arts." The bearers of this knowledge were scholars who usually came from the upper classes, but who themselves did not as a rule occupy any functions of direct rule. Rather, they formed a scholarly elite with a partly class-neutral recruitment.

The precarious status of this elite meant that the transmission of the new knowledge of this level was always at risk. Until medieval universities were founded, the intergenerational continuity of the social role of the scholar was never permanently secured. Scholars existed in social niches such as ancient academies, Christian monasteries or feudal courts. It is true that the body of knowledge was in principle permanently represented externally in manuscripts. However, as the fate of the ancient libraries shows, such representation could not prevent historical discontinuities in the transmission of knowledge. In particular, only a fraction of the knowledge of the ancient world was handed down to the Arab and later the Christian Middle Ages.

Yet such discontinuities primarily concerned the *content* of knowledge; less so the new types of knowledge *structures* associated with the independent production of knowledge, which were the result of the reflexive structuring of knowledge systems characteristic of this stage and closely related to its productive independence. This reflexive structuring, which is usually expressed by the term theory, has indeed survived all discontinuities in the transmission of knowledge qua content. As is particularly clear from the origin of Greek geometry in operations of geometric construction (Steele 1965), and the origin of Greek arithmetic in operations of practical arithmetic (Lefèvre 1981), the decisive step that led to theoretically structured knowledge systems consisted in the reflection of symbolic actions, whereby the written-language representation of the structural peculiarities and regularities of such actions formed an elementary prerequisite for this reflection (Klix 1993, Part II, Chap. 7).

It therefore seems anything but coincidental that the beginnings of theoretical mathematics and of logic (indirect proof and dialectics) coincide (Szabó 1968). Logic, which was elaborated in the philosophy of antiquity (Plato, Aristotle, Sextus Empiricus etc.), was a generalized theory of the possibilities of action using the symbol system of a written language. At this stage in the development of knowledge systems, it was still indistinguishable from a theory of natural language. With the new technique of definition, meanings of linguistic designations became the object of conscious structuring, and with proof, connections of content became the object of formal operations using linguistic forms.

In general, the new knowledge of this level is thus linguistically reflexively structured knowledge represented in symbol systems that are also reflexively structured by definitions, abstract terms, Euclidean constructions, deductive theories, etc. This knowledge system is based on theorizing the representation of knowledge. By thematizing the representation of knowledge in a logic, arithmetic, etc. that is not yet distinct from language theory at this level, these knowledge systems gained a metacognitive component.

The philosophical integration of bodies of knowledge made content-related contexts the object of formal operations using linguistic forms, and structured bodies of knowledge using logical operations (conceptual analyses, hierarchical links through conceptual trees, etc.). The limits of this structuring are set by the limits of language as a symbol system: where the conceptual meanings of reflexively constructed concepts were no longer related to the primary meanings of the reflected concepts, the emptiness of a merely formal structuring was unavoidable (scholasticism). As long as the structuring of knowledge was based only on the reflection of symbolic action, the structure remained external to the contents of experience. Insofar as knowledge of this level was theoretical knowledge, it was therefore not directly but only indirectly grounded in experience via language, and insofar as it directly re-inflected experience, it did not depend on the theoretical form. In this respect, the orientation function of the Aristotelian system of knowledge, which had paradigmatic significance for almost two thousand years, is also not to be understood primarily on the basis of its logical structuring. Rather, it was based on a fact that was not a consequence of the specific theoretical form, namely on the degree of objectivity that integrated knowledge had achieved in Greek antiquity.

Only in the rather sparse technological literature (Heron, Frontinus, Vitruvius, etc.), which did not belong to this new type of theoretical knowledge, even if it borrowed from it, can approaches be found to overcoming this gap between theoretical knowledge and experiential knowledge insofar as in it knowledge of production (engineering knowledge) was compiled, systematized, and handed down in writing for the first time, as was later to become characteristic of the theories of the modern era. Finally, Archimedes' writings on statics represent the unique case at this level that instruments of the engineer—in the form of "simple machines"—were theorized and made the basis of a mathematical-deductive representation of empirical knowledge.

Knowledge Systems in Modern Europe: The Systematic Production of Experiential Knowledge

The novel knowledge characteristic of modern Europe from the end of the Middle Ages to the turn of the nineteenth century is distinguished by the fact that it is systematically produced experiential knowledge. Such experiential knowledge characterizes not only the natural sciences (mechanics, physics, astronomy linked to mechanics, chemistry, geology, physiology, etc.) and engineering knowledge (from Alberti to Watt), but also knowledge systems of the social sciences such as economics. In connection with this development, integrations of the bodies of knowledge emerge, such as modern philosophical atomism, mechanicism and other philosophies of nature, which can be understood as scientific world views (immanent explanation of the world, banishment of final causes, etc.).

The basis for this new kind of knowledge was a changed relationship between traditional knowledge and production practice, for which the use of natural forces by means of technical apparatus and equipment (machines, chemical production processes, etc.) on a historically new scale was a decisive prerequisite. Not only did the theorization of engineering actually form an essential starting point for the emergence of the modern natural sciences (Grossmann 1935; Klein 1994; Klemm 1979; Lefèvre 1978; Wolff 1978; Zilsel 1976), but this process also developed the conditions for theoretical knowledge of nature itself to become eventually a factor in production.

This new relationship between knowledge and production also fundamentally changed social distribution of knowledge and the way it was communicated and transmitted. The printing press invented in the late Middle Ages acquired a key function for the new public sphere of knowledge that now became necessary. The roots of a general school system also lie in this development (Grundel 1928/1929). Special educational institutions also emerged, namely institutions for practitioners who had previously been excluded from the knowledge of the educated elite, where both theoretical and practical knowledge were transmitted. Eventually, academies were founded such as those in Florence in the sixteenth century or Gresham College, in whose tradition the various schools for special techniques (building academies, arithmetic schools, military academies, etc.) and teaching institutions such as the *École Polytechnique* were established. In this context, the scholarly language of Latin lost its overriding importance in favor of scientific and technical literature written in the respective vernacular language. Likewise, new non-university research institutions, scientific academies modelled on those in London and Paris, scientific societies, and scientifically curated natural history collections and botanical gardens were founded, which, like the *Muséum d'histoire naturelle* in Paris, developed into important centers of scientific research. The regularly published academy records marked the beginning of a new kind of publication system for scientific knowledge and technical developments, which found its characteristic form in modern scientific journals.

The importance that the new relationship between theory and production practices using natural forces had for scientific knowledge was demonstrated externally by the development of new scientific instruments (telescope, microscope, astronomical and geodetic surveying instruments, thermometer, barometer, etc.) as a means of acquiring knowledge. The emergence of these new instruments, however, only signaled a more profound change that centrally affected the process of cognition and the structuring of knowledge, the specificity of which could only be seen at the experiment, a change that was already recognized in its full scope by Francis Bacon (1620/2000).

Experimentation formed the core of the new method of systematic production of experiential knowledge. The scientific instruments used for these new experiments were largely based on modern technology. However, this technology was not only an external prerequisite for the production of such instruments, but the experiments were basically nothing other than arrangements of the technical apparatuses and installations characteristic of the new form of production, only that the processes determined by natural law and produced with them were not used productively in the experiment, but were carried out solely for the purpose of recognizing their lawfulness.

This is why theorizing the experiences of engineers, technicians, metallurgists, pharmaceutical chemists, etc., whose work was characterized by the use of such technical apparatuses and equipment, provoked the emergence of a new type of experimentally based theory formation, which is the hallmark of the knowledge systems of modern natural science. The experiment was a direct consequence, but not a prerequisite of this development: experimentation was a specific practice of gaining knowledge which supplemented the theorization of practice.

Just as in the natural sciences, experiences were theorized in the social sciences, for example in political economy. Although such sciences could not, by their very nature, draw on experiments to study the laws of economic action, they could make use of economic policy instruments such as trade balances and statistics, and could draw on experience gained from economic policy measures taken by governments (money supply policy, mercantilist trade policy).

Perhaps the most significant novelty of this stage is the fact that the means of labor—indeed the specific means of labor of the modern era—which implicate nature's modes of action in production independently of the physicality of human beings, were used as means of cognition, modified, and further developed in experimental designs for cognitive purposes. As important as the invention of new symbol systems and the new analytical methods based on them (analytical geometry, differential calculus, etc.) undoubtedly were, that they were now used as the characteristic means of development and theoretical representation in the exact natural sciences—i.e. the empirical sciences—is due to the fact that the quantities they relate and the nature of their relationships can be obtained by measuring them using technical or experimental means.

In general, therefore, the novel knowledge of this level is produced by an integration of theory and practical knowledge. It is obtained using scientific instruments to perform actions that aim to acquire knowledge about the possible results of these actions and to present them as empirically verifiable and falsifiable statements. The possibility offered by specifically modern production of using technology to perform natural processes constitutes the basis of this knowledge system. These possibilities are used in experiments to systematically produce experiences, which are represented in reflexively structured symbol systems in such a way that the reflected empirical experience organizes the logical operations within the body of theory, thus structuring the knowledge systems.

The achievement of integrating this reflexive structuring were, however, limited by the initially still narrow basis of knowledge that could be systematically acquired using the new means of cognition. As a result, the newly structured knowledge systems of this stage show a disproportion between the methodically guided acquisition of new individual insights and the natural philosophical (atomistic, mechanistic, dynamistic, etc.) attempts to give these new insights a common conceptual framework and to integrate them into an overarching scientific worldview. Through experimentation and methodically controlled observation, the statements that formed the basis of the modern knowledge systems gained an intersubjective binding force that had previously been almost inconceivable. In contrast, the reflexively constructed systems of natural philosophy, whose claims to explanatory power had to compete with the overall theories of the Aristotelian tradition and satisfy the demands of an emphatic, modern concept of reason (Kepler's world harmonics, Galileo's impetus theory, Newton's active principles, Leibniz's theory of monads, Kant's metaphysics of forces, Lamarck's transformation theory, etc.), were for a long time the subject of undecidable, principled disputes for their contemporaries—and rightly so, because from today's point of view they all appear to be reductionist generalizations and simplifications. With few exceptions, their basic assumptions later proved to be empirically untenable. It was only with the growth of methodically controlled accumulated knowledge that theories of empirical science also acquired intersubjective binding force. However, this process was accompanied by a fundamental change in the structure of modern knowledge systems, which is commonly referred to as "specialization" and which became the dominant characteristic of a new historical stage in the development of knowledge systems.

Knowledge Systems in the Industrial Age: Specialization and the Disciplinary Canonization of Knowledge

Around the beginning of the nineteenth century—at an uneven pace among the various disciplines of modern science—a differentiation of specialized fields of knowledge began, developing highly specific methods of acquiring knowledge, the adequate mastery of which sometimes required years of study. As a result of this development, a historically new gap emerged between general scientific knowledge and specialized knowledge, codified in artificial technical languages and complex, subject-specific symbol systems, which objectively embodies the rationality of the state of scientifically controlled knowledge achieved in the industrial age.

Institutionally, this development took the form of the emergence of modern scientific disciplines (Jungnickel and McCormmach 1986; Stichweh 1984) oriented towards the specific means and techniques which science now used for the systematic and methodically controlled acquisition of knowledge, and which constituted the objects of the disciplines. It led to an increasing differentiation of teaching and research, but at the same time also included an increasing differentiation of fields of knowledge and corresponding specializations within the individual disciplines, and led in particular to a separation between pure and applied fields of knowledge.

This specialization was also associated with an expansion of the subject areas covered by systematic empirical research. In particular, disciplines emerged for the first time that made knowledge itself the subject of systematic empirical research, namely social science disciplines (psychology, education, sociology, history of science, etc.), which differentiated themselves from the traditionally integrative humanities, especially philosophy, and—oriented towards the model of the natural sciences—developed their own methods of empirical knowledge acquisition.

On the other hand, specialization led to the collapse of the overarching theoretical integration of knowledge that had initially been characteristic of a modern science oriented towards natural philosophy. A science-immanent theoretical connection between bodies of knowledge was replaced by pragmatic considerations and evaluations that aligned the approaches and results of scientific research in relation to general socio-political objectives pursued through education and science policy, or to specific technological tasks through large-scale research management. On the one hand, integrating the findings of different disciplines for the purpose of decision-making practice in the respective scientific organizations was and is carried out by experts, but on the other hand, it occasionally also leads to the emergence of new types of metadisciplines (research on education and science management, etc.).

The background to this development is the fact that with the industrial revolution, knowledge changed its function, and the acquisition and transmission of knowledge gained an almost strategic importance for social reproduction. Specific, occupation-oriented education no longer seemed secure without expanding general basic knowledge. The modern civic institutions of general education, and occupation-specific school and post-school education with their own integrating knowledge systems, emerged (introduction of compulsory schooling, canonization of school knowledge in both primary school and grammar school education, upgrading of so-called comprehensive education, development of theoretically based, school-based vocational training, foundation of technical universities and colleges, reform of traditional universities, etc.). At the same time, institutions of higher education developed into state-supported institutions with integrated research and teaching, which were also joined by new, non-university research institutions.

This novel form of knowledge production had two fundamental structural consequences for the structure of knowledge systems. First, the close connection between the representation of knowledge in the specialized disciplines and the representation of knowledge in natural language was lost. Secondly, the form of representation of knowledge in the specialized disciplines was depersonalized, and the structure of knowledge systems became intersubjectively binding to an unprecedented degree.

Already in the previous stage, the terminology in which the novel knowledge of the empirical sciences was formulated was so marked by the technical apparatuses and facilities of knowledge acquisition that the connection to natural language, the communication system for generally accessible knowledge, became metaphorical. However, it was precisely this metaphorical approach that opened up access to the specialized and precise concepts of the theoretically structured bodies of knowledge. With the emergence of the specialized disciplines in which the actual production of knowledge takes place in the industrial age, even this metaphorical connection to natural language is largely lost, and the communication of expert knowledge becomes a problem that can only be solved pragmatically on a case-by-case basis. While nineteenth-century attempts to make expert knowledge accessible through "popularization" represented serious efforts to uphold the claim to a scientific worldview, in the twentieth century the popular representations of advanced knowledge degenerated into subjective translations that no longer open up access to the inner rationality of knowledge. The flip side of the promise of the popularization of science is the disconnection of specialized knowledge from the living world, which has resulted in an unacknowledged crisis of scientific education that has not been overcome to this day.

In the course of this development, the internal rationality of knowledge changes its character insofar as the binding nature of scientific knowledge, which originally applied only to empirically secured individual statements (in physics: mechanical laws, Kepler's laws, etc.), is now also increasingly achieved for the conceptual foundation and theoretical structuring of knowledge systems, whereby natural philosophical systems have been replaced by the area-specific theories of the individual disciplines.

This development was caused by the extraordinary broadening of the knowledge base for the formation of specialized scientific theories, which—at least in the natural sciences, or more precisely, in those sciences that had become a condition for social reproduction—made possible a new quality of appropriate theoretical structuring of knowledge. It appeared outwardly as an increase in the stability of theoretical structuring vis-à-vis new individual knowledge and made it possible to cumulatively expand knowledge in differentiated knowledge systems that could no longer be fundamentally endangered even by "scientific revolutions" (Kuhn 1962). This development was achieved through the novel organizational form of scientific knowledge acquisition that emerged in the nineteenth century and, in particular, through the changed role of scientific communication in the construction of scientific theories: in specialized disciplines, the integration of individual results is no longer primarily the achievement of the individual researcher, but the result of communication within the discipline.

In general, it can be said for this stage that scientific knowledge, especially the knowledge of the natural sciences, engineering and socio-technological disciplines, has become a determining factor in production. It is produced, distributed and passed on in a complex institutional system in the form of specialized knowledge. As with the knowledge of the previous stage of development, this knowledge is reflexively structured, but is now represented by empirically founded symbol systems whose basic structures have become stable for the specialized sub-areas of human knowledge, which is gained and passed on in specialized disciplines with their own objectified conceptual and theoretical structures.

Knowledge Systems in the Computer Age: A New Stage of Development?

In the recent past, the development of information technology has led to some fundamental changes in the representation of knowledge, which again point to the emergence of new types of knowledge system structures. Although electronic data processing initially did little to change the dominance of written knowledge representation, this technology for the first time opens up the possibility of shifting the operative functions of human thinking and dealing with written knowledge into the medium of representation itself.

This combination of knowledge content with the operational functions of human thought (Newell and Simon 1972) seems to be a general feature of the present and future development of new forms of knowledge representation based on information technology (Crane 1991; Findler 1979). However, attempts to simulate natural thought processes have shown that cognitive psychology abstracting from knowledge contents on the one hand and formal logic on the other hand have expressed these operative functions only very insufficiently in theoretical terms. New fields of research have therefore emerged in connection with the attempt to realize artificial intelligence, in particular the psychology of knowledge (Mandl and Spada 1988) and cognitive science (Johnson-Laird and Wason 1977; Minsky 1975, 1985), which have raised the traditional questions of logic in a completely new way and have also provided new answers by constructing extended and alternative logics (Besnard 1989; Collins and Michalski 1989), some of which are of a subject-specific nature (McDermott 1982). Cognitive science is characterized by an extended, transdisciplinary reflexivity with regard to the concept of knowledge, as was already characteristic of the Piagetian epistemological research program. Influenced by new techniques of knowledge processing, a kind of psychological-empirical theory of knowledge has emerged that addresses the question of the nature of knowledge in its various forms and representations as well as questions such as the relationship between expert and novice knowledge or between everyday and scientific knowledge.

Even if the possibilities of "intelligent" machine processing of knowledge are judged more cautiously today than they were a few years ago, there are signs that the scientific discussion of the structure and function of knowledge systems will have far-reaching consequences for handling knowledge stocks in the long term (Johnson-Laird 1993). It is already apparent today that the increased speed of access to information alone is profoundly changing the way we deal with knowledge and at the same time placing new demands on the structuring of knowledge stocks. Indeed, information technology also makes it possible to codify the structure of represented knowledge differently from the linear form of spoken and written language, which has been characteristic of the entire written tradition up to now. Representation can be functionally adapted to the structures of the represented knowledge systems, a novel possibility already extensively used in databases and hypertexts (Barret 1989; Cunningham and Hubbold 1992; Shneiderman and Kearsley 1989). Whether the futuristic projection that the knowledge systems of the present will grow together in the near future into a single, technically realized, worldwide network of human knowledge is to be taken seriously as a real possibility must be left open here.

Universals of the Historical Development of Knowledge Systems

Changes in the contents, functions and structures of knowledge systems that can be observed in the historical development of social knowledge, raise two fundamental questions, the answers to which belong to the central problems of comparative psychology and especially of the psychology of knowledge: namely the question of cognitive universals and that of the psychological specificity of culture-specific forms of cognition. Are there psychological factors that remain unchanged in the process of the historical development of knowledge systems, and in which psychological factors are the historical changes of knowledge systems expressed?

The question of cognitive universals is closely related to the question of universal anthropological constants. As a biologically determined being, humans are endowed with specific abilities, such as the ability to acquire and use language, which all members of the human species possess regardless of cultural and historical differences. Such abilities form a basis of cultural development, but did not themselves arise in the process of cultural development. However, the question of cognitive universals here is not about the existence of universals at all, but whether there are also universals that are of an original psychological nature, especially those that play a decisive role in the acquisition, storage, retrieval, application and change and development of knowledge (Berry et al. 1988; Brown 1991; Dasen 1977; Lloyd and Gay 1981; Lonner 1980).

The problem of the psychology of knowledge can be narrowed down even more precisely. On the one hand, the content of knowledge itself can be affected by such universals, namely when certain concepts and thought structures do not result solely from learning processes, but are already genetically or epigenetically preformed in their basic structure (space, time, causality, etc.). Secondly, the cognitive functions that both enable and limit the mental processing of knowledge can be biologically determined in such a way that not only the quantitative performance but also forms of knowledge structuring and processing can be explained by this biological prerequisite (use of symbols, structure of memory, imagination, deductivity etc.). Finally, individual cognitive development can also be biologically shaped in such a way that the development of knowledge in the course of an individual's biography cannot be explained by learning processes alone, but at the same time passes through a sequence of universal ontogenetic developmental stages (development of sensory-motor coordination, language, spatial imagination, abstract thinking, etc.).

Of course, such problems were not first addressed with the emergence of the psychology of knowledge. Thus, the determination of human knowledge by fundamental "categories" (being, essence, quality, quantity, etc.) and by the "faculties" of a thinking psyche (memory, imagination, thinking, moral judgement, etc.) has been the subject of theoretical reflection since antiquity in the philosophical tradition of thinking about the nature of knowledge. Until the beginning of the modern era, a universalistic metaphysics prevailed, which regarded the fundamental categories canonized in the tradition of Plato and Aristotle as universal properties of objects (ontology).

With the ideological disputes associated with the emergence of modern science, new fundamental categories with an experiential foundation but a similar universal claim were set against this ontology. At the latest, universalism itself became in need of justification. Thus, in particular in the rationalist tradition, which had a long lasting effect, especially through the work of Kant, basic categories of classical physics (space, time, matter, motion, causality, etc.) were seen as "a priori" knowledge, which precedes all experience and is subject to neither historical nor cultural changes. Their justification as principles of thought or as products of reflection shifted ontology and thus the source of universalism to the cognizing subject, with the consequence that the radical alternative of regarding even such categories as relative, as determined by the contents of knowledge, as empiricism had momentously advocated, could no longer be dismissed as a possibility of thought. Finally, in the historicism of the nineteenth century, the possibility was even raised that the process of cultural development itself determined the basic categories that structured the knowledge of the respective epoch.

This was the problem when, with the establishment of the modern humanities and social science disciplines, the question of universal structures and functions of knowledge systems took an empirical turn (Wundt 1900–1920) and was finally made the subject of comparative cultural studies using the methods of empirical cognitive science, whereby special importance was attached from the outset to comparing cognition in recent oral cultures with cognition in the Western European cultural tradition. Initially, it was primarily the reports of explorers, missionaries and colonial officials that allowed the question of the cultural dependence of cognition to be posed in a new way. Such reports were naturally predominantly written in a form that made the cognitive achievements of primitive peoples appear as deficient and "primitive" when measured against our own knowledge systems. At the latest with Lévy-Bruhl's provocative description of the "mental world of the primitives" as prelogical (1910, 1922), the question of whether it was justified to interpret the thinking of pre-modern cultures simply as a primitive preliminary stage of thinking in modern culture became a central topic of debate. Gestalt psychology in particular made the question of the inherent laws of cognition at early stages of development the subject of psychological research early on through studies of both animal cognition (Köhler 1921) and the cognition of non-European cultures (Wertheimer 1925). Using the example of the concept of numbers, Wertheimer outlined a research program to empirically resolve the question of the nature of thought in pre-modern societies:

> One approaches the thinking of the so-called primitive peoples with our categories—number, cause, abstract concepts—as they have developed in our biological, social conditions, in our history, with the cogent prejudice that their thinking is nothing but a preliminary stage, only vaguer, rudimentary, possibly more incompetent; and thereby blocks oneself the way to a real knowledge of what is actually given. … It is a matter of clearing the ground for research; of accessing the concrete problems which do not arise in dogmatic European consideration, or which arise in an oblique way; first of all: to indicate factors which come into question and their functional relationships. (Wertheimer 1925, 107)

Turning away from a projection of modern categories onto thinking in the context of cultures that differ substantially from modern culture was identified as the most important prerequisite for being able to adequately pursue the specifics of such thinking. The question of universalism became an open problem, and there was even a tendency towards culturally relativistic interpretations at first, since valid, overarching descriptive categories were lacking. It was not until psychological structuralism (Piaget 1950) that a theory emerged within psychology that took up the humanistic-philosophical tradition again and consistently understood the cultural differences in the structure and function of knowledge systems as developmental stages of an epigenetically determined cognition with universal, mathematical-logical structures. The universalism of such a structuring of all knowledge (for example, through categories such as quantity, order, space, time, etc.) is justified here with the assumption that such structures emerge as reflective abstractions from ultimately biologically determined coordination of action. According to this theory, the structures of knowledge are therefore subject to ontogenetic as well as evolutionary development processes, without their nature changing in principle. According to this theory, knowledge systems have universal structures which also appear in linguistic representation, but which do not have a linguistic origin, and are predetermined in all cultures that structure knowledge linguistically by a universal context of action.

In linguistics, too, where, in contrast, the assumption dominates that knowledge structures are primarily or even exclusively determined by the structures of language, the results of investigation into the structures of language in different cultures in the wake of Whorf (1956) were initially interpreted emphatically as evidence for the relativism of thought structures—a relativism that found a paradigmatic expression in the so-called "Sapir-Whorf hypothesis" (Pinxten 1976). The extreme opposite position is represented by the tradition going back to Chomsky (1965, 1957), in which universal structures of thought are interpreted directly as biologically determined, generative "deep structures" of language, structures which, unlike the universal structures of psychological constructivism, are not subject to developments that could lead to historical changes in the structuring of knowledge systems (Piattelli-Palmerini 1980).

The controversy thus structurally created in both psychology and linguistics by the problem of universalism has given rise to and continuously stimulated diverse research on numerous aspects of the problem. The question of universal cognitive structures, functions or ontogenetic developmental processes, which is crucial for an assessment of historical changes in knowledge systems, still remains essentially unanswered (Pinxten 1976; van de Vijver and Hutschemaekers 1990). On the one hand, this is due to the considerable difficulties in conducting psychological studies in the few still existing cultures without Western-European cultural influence, which correspond to the methodological requirements of empirical psychological research; on the other hand, such studies are also necessarily limited in their subjects. For example, intercultural performance comparisons predominate in comparative cultural research; in contrast, developmental psychology studies that could illuminate the extent and manner of cultural influence on developmental processes are largely lacking.

The problem becomes particularly apparent in the area in which the rationalist tradition has long assumed an "a priori" knowledge and which has therefore also become the subject of comparative cultural studies of the most diverse kinds, namely in the area of basic categories of the comprehension of nature (space, time, number, matter, movement, causality, etc.) and of logical reasoning. The question of which structures of such concepts are culture-independent is still the subject of extraordinarily controversial discussion; cf. for example, on the concept of space: Ellen and Thinus-Blanc (1987), Landau and Jackendoff (1993), Levinson (1991, 1992), Liben et al. (1981), Pick and Acredolo (1983), Pinxten et al. (1983); on the concept of time: Block (1990), Dux (1992), Elias (1984), Friedman (1982); Gardet et al.

(1976), Levin and Zakay (1989); on the concept of number: Ashcraft (1992), Boysen and Capaldi (1993), Damerow (1993a, 1993b), Ginsburg (1983), Hurford (1975, 1987) and on logical reasoning: Bobrow (1984), Hutchins (1980), Langer (1980, 1986), Luria (1976), Scribner (1975). Interpreting the historical change to which these concepts are subject is directly affected by the insufficient clarification of their nature. As long as comparative cultural psychology does not succeed in identifying the anthropological constants in such changes, any interpretation of the historical development of these categories from the situation-bound cognitive structures of the scriptless period via the structures of Aristotelian categories closely linked to those of natural language, to the basic concepts of classical physics and those of modern physics, is beset by the problem that the boundary between what must be explained psychologically and what must be explained historically cannot be precisely determined. On the other hand, precisely because of the close connection between the psychological question of cognitive universals and the historical question of the development of such categories, it is also possible for the psychology of knowledge to make use of the results of historical research in order to solve this problem, which transcends the disciplines.

The Psychological Dimension of the Historical Development of Knowledge Systems

The second question, closely related to the historical changes in the contents, functions and structures of social knowledge, is the question of the psychological characteristics of cognition at the various stages of historical development. To what historical changes are the psychological processes that play a role in the mental processing of knowledge systems subject?

It is obvious and factually justified to describe cognition at the historical stages of cognitive development and, in particular, to describe the characteristic cognitive performances whose first appearance in each case signals the attainment of a new stage, using categories of contemporary psychological research. The categories used to describe cognitive performance at the various stages of ontogenetic development can also be used to describe the cognitive functions and structures in other historical periods. However, this inevitably leads to the fact that the historical sequence of increasingly developed cognitive performances shows a certain parallelism with the ontogenetic developmental process, because in both cases the cognitive performances that mark the attainment of a new stage represent more elementary cognitive functions in the early developmental stages than in the later ones. To the extent that such cognitive performances are universal in nature, but the respective historical circumstances are only external and contingent conditions of their epigenetic unfolding, this parallelism also has a theoretical significance. The historical development of cognitive performance is then only a result of the same epigenetic process that also determines the ontogenesis of basic cognitive functions and structures. However, to the extent that such cognitive performance is culture-specific in nature, ontogenetic development must conversely be interpreted as the reconstruction of culturally represented cognitive structures. The problem of universalism is therefore significant in a further sense for the interpretation of the historical development of basic cognitive categories than the one described above, namely for determining the relationship between the developmental laws of ontogenetic and those of the historical development of knowledge systems.

Normally, however, such implications of psychological investigations remain unspoken. Here, Piaget's theory of the psychogenesis of basic concepts of scientific thought deserves special attention, because Piaget is one of the few psychologists to have concretely explicated the implicit historical-theoretical consequences of assumptions of psychological universalism (Piaget 1950, 1978; Piaget and Garcia 1989).

If one follows Piaget's interpretation of the results of his developmental psychological studies, then there is a fixed canon of universal logical and mathematical structures of cognition, which in particular form the historically unchanging basis of arithmetic, geometry, and physics. If these cognitive structures could already be demonstrated for the earliest stages of historical development, then their epigenesis would be a specific phenomenon of ontogenetic development. Now, however, at least for some of the recent oral cultures—whose thinking Piaget characterizes as "prelogical"—it can be taken as certain that these structures are not developed or are only rudimentary. Consequently, Piaget concludes that the historical development of arithmetical, geometrical, and physical concepts must first have passed through a phase of development in strict parallelism to ontogenetic development, up to the point at which these concepts are formed, i.e. up to a stage that essentially corresponds to the everyday thinking of an adult in modern culture. The further historical development of mathematical and physical knowledge is—if one follows Piaget's theory—based on this universal foundation, but in contrast to the preceding, epigenetically determined phase, it is open in its outcome. When the turning point from the prehistory to this actual history of the knowledge structures concerned was reached can be determined historically. According to this theory, the turning point for the concept of number was already reached in the time before the early high cultures, but for the physical concept of force it was only reached in the time of Newton.

These consequences that Piaget drew from his assumptions of universalism lack historical plausibility. One can only conclude that these assumptions are flawed and must be brought into a form that is in harmony with historical development (Damerow 1993b). However, if one disregards this shortcoming of Piaget's theory, it is also clear that, with regard to historical development, a number of original psychological questions arise concerning the acquisition, storage, application, and transformation of knowledge, which historical research alone is not able to answer. In this situation, historians usually make do with plausibility considerations, which, however, often do not stand up to serious psychological scrutiny. A few examples may explain the historical task that arises for the psychology of knowledge.

In the case of knowledge systems in prehistoric oral cultures, the difficulties of determining the functions and structures of these systems are mainly due to the lack of information about mental processes in the time before the invention of writing.

Consistently applying the knowledge gained by comparative psychology about thinking in still existing oral cultures to the archaeological evidence of already intellectually shaped, pre-written cultures would contribute significantly to replacing the arbitrariness of the mostly anachronistic speculations about the cognitive performances to be presupposed by well-founded hypotheses. The existing approaches in this direction, for example, attempts at psychological interpretation of Paleolithic graphic and symbolic activities (David et al. 1988; Davidson and Noble 1989; Lewis-Williams and Dowson 1988) or the attempts to relate Piaget's theory to ethnological findings (Hallpike 1979), have already led to a promising objectification of the discussion about the biological and historical origins of cognitive performance.

From the period of knowledge systems in early written cultures onwards, written documents provide information about existing knowledge and the way of dealing with knowledge stocks. Special attention should be paid to the new kind of expert knowledge, which can be found for the first time in this period. One of the obvious problems is that knowledge in this period is still closely related to certain types of action, without knowledge of which the cognitive functions of this knowledge can hardly be determined, but about which sufficient information is rarely available. For example, the various forms of "pre-Greek mathematics" in the early written cultures (Babylonia, China, Egypt, Central America) have been handed down to us through numerous sources whose mathematical content can generally be reconstructed in modern terms without too much difficulty, but whose heuristics seem so strange that the meaning of the documented operations remains obscure. Therefore, one is often content with reproducing the structure of the texts in modern terminology and interpreting them in modern mathematical categories, although the program of reconstructing expert knowledge in its own terms was formulated early on, especially on the part of philology; for the Babylonian culture, for example, by Landsberger (1926). In contrast, the question of what knowledge the documented solutions to mathematical problems are based on and with what representations and what cognitive operations they may have been achieved has rarely been investigated. However, it is precisely with such investigations that the most interesting results for understanding pre-Greek mathematical texts have been achieved in recent times, whose importance for the question of the nature of deductivity can hardly be overestimated; cf. for example on recent primitive peoples in the transition to written culture: Scribner and Cole (1981); for Babylonian mathematics: Høyrup (1990); for Chinese mathematics: Chemla (1991).

The linguistically reflexively structured knowledge of the knowledge systems of cultures with independent knowledge production represents a further challenge of the historical tradition to the psychology of knowledge. The structures observable for the first time in the documents of this period (dialogical form of representation, ontology, deductive order, definition of meaning, reflection on linguistic representation, etc.) are so closely linked to familiar psychological processes (abstraction, generalization, conceptualization, reflection, etc.) (Nelson 1992) that understanding their emergence in a particular period of the historical development of knowledge systems would contribute equally to clarifying the psychological nature of processes constituting the deductive structure of modern theories, and to clarifying their historical origin.

With the knowledge systems of European modernity, the targeted, experience-led construction of theoretical models is added as a new moment in the development of knowledge systems. In this period of modern thought, the basic categories of today's scientific and social world view were created through the construction of mental models, which probably had their origins in everyday thinking, but now had to integrate new, scientifically acquired experiences (Damerow et al. 1992). The strong individualization associated with overcoming the canonical theoretical forms of the Middle Ages in this period offers special possibilities for explaining the cognitive achievements associated with this development psychologically. Nevertheless, intensive research in the history of science into the historical details of this development faces a challenge in explaining the problem-solving processes and modern strategies on which the drastic change in thinking in the period after the Copernican turn is based.

Finally, the emergence of specialized expert knowledge within the framework of objectified conceptual and theoretical structures, which has become the hallmark of knowledge systems in the industrial age, is particularly topical in that the problems of acquiring, communicating and integrating expert knowledge that have arisen from this development have not been overcome to this day and have led to separate research directions in the psychology of knowledge (expertise research, conceptualization and development of expert systems, etc.). Explaining the conditions under which knowledge systems have assumed the structure of modern expert knowledge and the function of a means for highly specialized problem-solving processes in a form of knowledge production system subject to a high degree of division of labor therefore contributes directly to solving the problems for which the psychology of knowledge has emerged as a separate branch of psychology.

If one compares even this cursory catalogue of problems with the overview given at the beginning of psychological studies that have the historical development of cognition as their subject in one way or another, or that take cognition in other historical epochs into account at least to a certain extent, it becomes clear that there is a glaring disproportion between the meticulous reconstruction of real history by historians and the only very eclectic approaches to reconstructing the history of knowledge systems and their fundamentally changing historical structures and functions by psychologists. The small number of such studies and the few indications that can be gained from them for answering the questions concerning the historical change of knowledge systems indicate that there is a marked deficit in psychological research.

References

Aebli, H. 1981. *Denken: das Ordnen des Tuns* (2 vols.). Stuttgart: Klett-Cotta.
Alt, R. 1956. *Vorlesungen über die Erziehung auf frühen Stufen der Menschheitsentwicklung*. Berlin: Volk und Wissen.
Ascher, M., and R. Ascher. 1972. Numbers and relations from ancient Andean Quipus. *Archive for History of Exact Sciences* 8 (4): 288–320.
Ashcraft, M. H. 1992. Cognitive arithmetic: A review of data and theory. *Cognition* 44 (1–2): 75–106.
Bacon, F. 2000. *The new Organon* (1620), eds. L. Jardine & M. Silverthorne. Cambridge: Cambridge University Press.
Barret, E., ed. 1989. *The society of text: Hypertext, hypermedia and social construction of information*. Cambridge, MA: MIT Press.
Bell-Krannhals, I. 1990. *Haben um zu geben. Eigentum und Besitz auf den Trobriand-Inseln, Papua New Guinea*. Basel: Wepf.
Berger, P. L., and T. Luckmann. 1966. *The social construction of reality: A treatise in the sociology of knowledge*. Garden City, NY: Anchor Books.
Bernal, J. D. 1939. *The social function of science*. London: Routledge.
Berry, J. W., S. H. Irvine, and E. B. Hunt, eds. 1988. *Indigenous cognition: Functioning in cultural context*, NATO ASI Series. Vol. D 41. Dordrecht: Nijhoff.
Besnard, P. 1989. *An introduction to default logic*. Berlin: Springer.
Block, R. A., ed. 1990. *Cognitive models of psychological time*. Hillsdale: Erlbaum.
Bobrow, D. G., ed. 1984. *Qualitative reasoning about physical systems*. Amsterdam: North-Holland.
Boehmer, R. M., P. Damerow, R. K. Englund, and H. J. Nissen, eds. 1991/1993. *Materialien zu den frühen Schriftzeugnissen des Vorderen Orients (MSVO) [2 Bde., Reihe wird fortgesetzt]*. Berlin: Gebr. Mann.
Born, M. 1965. *Von der Verantwortung des Naturwissenschaftlers*. Munich: Nymphenburger Verlagshandlung.
Boysen, S. T., and E. J. Capaldi, eds. 1993. *The development of numerical competence: Animal and human models*. Hillsdale: Erlbaum.
Brainerd, C. J. 1979. *The origins of the number concept*. New York: Praeger.
Brown, D. E. 1991. *Human universals*. Philadelphia: Temple University Press.
Bruner, J. S., R. R. Olver, P. M. Greenfield, et al. 1966. *Studies in cognitive growth*. New York: Wiley.
Chadwick, J. 1958. *The decipherment of linear B*. Cambridge: Cambridge University Press.
Chemla, K. 1991. Theoretical aspects of the chinese algorithmic tradition (first to third century). *Historia Mathematica* 42:75–98.

W. Lefèvre, *Theory of Nature and Mode of Production*, Political Epistemology,
https://doi.org/10.1007/978-3-032-02295-0

Chomsky, N. 1957. *Syntactic structures*. The Hague: Mouton.

Chomsky, N. 1965. *Aspects of the theory of syntax*. Cambridge, MA: MIT Press.

Clark, G. 1970. *Science and Social Welfare in the age of Newton*. Oxford: Clarendon Press.

Clement, J. 1983. A conceptual model discussed by Galileo and used intuitively by physics students. In *Mental models*, ed. D. Gentner and A. L. Stevens, 325–339. Hillsdale: Erlbaum.

Cole, M., and S. Scribner. 1974. *Culture and thought: A psychological instruction*. New York: Wiley.

Collins, A., and R. Michalski. 1989. The logic of plausible reasoning: A core theory. *Cognitive Science* 13:1–49.

Crane, G. 1991. Composing culture: The authoring of an electronic text. *Current Anthropology* 32 (3): 293–302.

Cunningham, S., and R. J. Hubbold, eds. 1992. *Interactive learning through visualization: The impact of computer graphics in education*. Berlin: Springer.

Damerow, P. 1993a. Voriiberlegungen zu einer historischen Epistemologie der Zahlbegriffsentwicklung. In *Der Prozeß der Geistesgeschichte: Studien zur ontogenetischen und historischen Entwicklungslogik des Geistes*, ed. G. Dux, 248–322. Frankfurt am Main: Suhrkamp.

Damerow, P. 1993b. Zum Verhältnis von Ontogenese und Historiogenese des Zahlbegriffs. In *Die Konstruktion kognitiver Strukturen: Perspektiven einer konstruktivistischen Entwicklungspsychologie*, ed. W. Edelstein and S. Hoppe-Graff, 195–259. Bern: Huber.

Damerow, P., and R. K. Englund. 1989. *The Proto-Elamite Texts from Tepe Yahya. With an introduction by C. C. Lamberg-Karlovsky*, American School of Prehistoric Research Bulletin. Vol. 39. Cambridge: Harvard University Press.

Damerow, P., G. Freudenthal, P. McLaughlin, and J. Renn. 1992. *Exploring the limits of preclassical mechanics. A study of conceptual development in early modern science: Free fall and compounded motion in the work of Descartes, Galileo, and Beeckman*. New York: Springer.

Damerow, P., and W. Lefèvre, eds. 1981. *Rechenstein, Experiment, Sprache. Historische Fallstudien zur Entstehung der exakten Wissenschaften*. Stuttgart: Klett-Cotta.

Dasen, P. R. 1977. Are cognitive processes universal? A contribution to cross-cultural piagetian psychology. In *Studies in cross-cultural psychology*, ed. N. Warren, vol. 1, 155–201. London: Academic Press.

David, N., J. Sterner, and K. Gavua. 1988. Why pots are decorated. *Current Anthropology* 29:365–389.

Davidson, I., and W. Noble. 1989. The archaeology of perception: Traces of depiction and language. *Current Anthropology* 30:125–155.

Dehaene, S. 1992. Varieties of numerical abilities. *Cognition* 44:1–42.

Diamond, J. 1966. Zoological classification system of a primitive people. *Science* 151:1102–1104.

Diederich, W., ed. 1974. *Theorien der Wissenschaftsgeschichte. Beiträge zur diachronen Wissenschaftstheorie*. Frankfurt am Main: Suhrkamp.

Dröber, W. 1964. *Kartografie bei den Naturvölkern (Reprint der Originalausgabe von 1903)*. Amsterdam: Meridian.

Dux, G. 1992. *Die Zeit in der Geschichte: Ihre Entwicklungslogik vom Mythos zur Weltzeit*. Frankfurt am Main: Suhrkamp.

Elias, N. 1984. *Über die Zeit* (ed. M. Schröter). Frankfurt am Main: Suhrkamp.

Ellen, P., and C. Thinus-Blanc, eds. 1987. *Cognitive processes and spatial orientation in animal and man*, NATO ASI Series. Vol. D 36. Dordrecht: Nijhoff.

Engelkamp, J., and T. Pechmann, eds. 1993. *Mentale Repräsentation*. Bern: Huber.

Englund, R. K. 1990. *Organisation und Verwaltung der Ur III-Fischerei*. Berlin: Reimer.

Findler, N. V., ed. 1979. *Associative networks: Representation and use of knowledge by computers*. New York: Academic Press.

Freud, S. 1974. Civilization and its discontents [1929]. In *The future of an illusion, civilization and its discontents, and other works, 1927–1931, Volume 21 of The standard edition of the complete psychological works of Sigmund Freud*, ed. James Strachey, 57–145. London: Hogarth Press.

Friedman, W. J., ed. 1982. *The developmental psychology of time*. New York: Academic Press.

Gardet, L., A. J. Gurevich, A. Kagame, C. Larre, G. E. R. Lloyd, A. Neher, R. Panikkar, G. Pàttaro, and P. Ricoeur. 1976. *Cultures and time*. Paris: UNESCO Press.

Gelb, I. J. 1958. *Von der Keilschrift zum Alphabet: Grundlagen einer Schriftwissenschaft*. Stuttgart: Kohlhammer.

Gentner, D., and A. L. Stevens, eds. 1983. *Mental models*. Hillsdale: Erlbaum.

Giere, R. N. 1988. *Explaining science: A cognitive approach*. Chicago: University of Chicago Press.

Giere, R. N. 1991. *Understanding scientific reasoning*. 3rd ed. Fort Worth: Harcourt Brace Jovanovich.

Giere, R. N., ed. 1992. *Cognitive models of science*. 3rd ed. Minneapolis: University of Minnesota Press.

Ginsburg, H. P., ed. 1983. *The development of mathematical thinking*. New York: Academic Press.

Gladwin, T. 1970. *East is a big bird: Navigation and logic on Puluwat Atoll*. Cambridge, MA: Harvard University Press.

Goody, J. 1986. *The logic of writing and the organization of society*. Cambridge: Cambridge University Press.

Goody, J. 1987. *The interface between the written and the oral*. Cambridge: Cambridge University Press.

Gould, S. J. 1980. A Quahog is a Quahog. In *The Panda's Thumb: More reflections in natural history*, ed. S. J. Gould, 204–213. New York: Norton.

Griese, A., and H. Laitko, eds. 1969. *Weltanschauung und Methode*. Berlin: VEB Deutscher Verlag der Wissenschaften.

Grossmann, H. 1935. Die gesellschaftlichen Grundlagen der mechanistischen Philosophie und die Manufaktur. *Zeitschrift für Sozialforschung* 4:161–230.

Gruber, H. 1981. *Darwin on man: A psychological study of scientific creativity*. 2nd ed. Chicago: University of Chicago Press.

Grundel, F. 1928/1929. *Die Mathematik an den deutschen höheren Schulen (Beihefte zur Zeitschrift für mathematischen und naturwissenschaftlichen Unterricht)*, ed. W. Lietzmann & W. Hillers, vols. 12/13. Leipzig: Teubner.

Gündel, R., ed. 1968. *Wissenschaft im Klassenkampf (Autorenkollektiv der TU Dresden)*. Berlin: Dietz Verlag.

Hallpike, C. R. 1979. *The foundations of primitive thought*. Oxford: Clarendon Press.

Harig, G., and H. Neels, eds. 1963. *Die Entwicklung der Wissenschaft zur unmittelbaren Produktivkraft*. Leipzig: Karl-Marx-Universität.

Harig, G., and J. Schleifstein, eds. 1960. *Naturwissenschaft und Philosophie Beiträge zum Internationalen Symposium über Naturwissenschaft und Philosophie anläßlich der 550-Jahr-Feier der Karl-Marx-Universität Leipzig*. Berlin: De Gruyter.

Hiepko, P., and W. Schiefenhövel. 1987. *Mensch und Pflanze: Ergebnisse ethnotaxonomischer und ethnobotanischer Untersuchungen bei den Eipo, zentrales Bergland von Irian Jaya (West-Neuguinea), Indonesien*. Berlin: Reimer.

Hiller, S., and O. Panagl. 1976. *Die frühgriechischen Texte aus Mykenischer Zeit: Zur Erforschung der Linear B-Tafeln*. Darmstadt: Wissenschaftliche Buchgesellschaft.

Horkheimer, M. 1937. Traditionelle und kritische Theorie. *Zeitschrift für Sozialforschung* 6 (2): 245–294.

Horkheimer, M. 1988. Traditionelle und kritische Theorie (1937). In *Horkheimer, Gesammelte Schriften, Band 4: Schriften 1936-1941*, ed. Alfred Schmidt, 162–225. Frankfurt am Main: Fischer Taschenbuch Verlag.

Horkheimer, M. 1992. Traditional and critical theory. Trans. Matthew J. O'Connell. In *Critical theory: Selected essays*, ed. M. Horkheimer, 188–243. New York: Continuum.

Hörz, H. 1976. *Marxistische Philosophie und Naturwissenschaften*. Berlin: Akademie-Verlag.

Høyrup, J. 1980. Influences of institutionalized mathematics teaching on the development and organization of mathematical thought in pre-modern times. In *Materialien und Studien*, vol. 20, 7–137. Bielefeld: Institut für Didaktik der Mathematik.

Høyrup, J. 1990. Algebra and Naive Geometry. *Altorientalische Forschungen* 17:27–69, 262–354.

Hurford, J. R. 1975. *The linguistic theory of numerals*. Cambridge: Cambridge University Press.

Hurford, J. R. 1987. *Language and number*. Oxford: Basil Blackwell.

Husserl, E. 1954. *Die Krisis der europäischen Wissenschaften und die transzendentale Phänomenologie*, ed. Walter Biemel. The Hague: Nijhoff

Husserl, E. 1970. *The crisis of European sciences and transcendental phenomenology*, trans. D. Carr. Evanston: Northwestern University Press

Hutchins, E. 1980. *Culture and inference: A Trobriand case study*, Cognitive science series. Vol. 2. Cambridge, MA: Harvard University Press.

Hutchins, E. 1983. Understanding micronesian navigation. In *Mental models*, ed. D. Gentner and A. L. Stevens, 191–225. Hillsdale: Erlbaum.

Janich, P., F. Kambartel, and J. Mittelstraß. 1974. *Wissenschaftstheorie als Wissenschaftskritik*. Frankfurt am Main: Aspekte Verlag.

Johnson-Laird, P. N. 1993. *Human and machine thinking*. Hillsdale: Erlbaum.

Johnson-Laird, P. N., and P. C. Wason, eds. 1977. *Thinking: Readings in cognitive science*. Cambridge: Cambridge University Press.

Jungnickel, C., and R. McCormmach. 1986. *Intellectual mastery of nature: Theoretical physics from Ohm to Einstein*. Chicago: University of Chicago Press.

Kambartel, F., and J. Mittelstrass, eds. 1973. *Zum normativen Fundament der Wissenschaft*. Frankfurt am Main: Athenäum-Verlag.

Klein, U. 1994. *Verbindung und Affinität - Die Grundlegung der neuzeitlichen Chemie an der Wende vom 17. zum 18. Jahrhundert*. Basel: Birkhäuser.

Klemm, F. 1979. *Zur Kulturgeschichte der Technik: Aufsätze und Vorträge 1954-1978*. Munich: Deutsches Museum.

Klix, F. 1992. *Die Natur des Verstandes*. Göttingen: Hogrefe.

Klix, F. 1993. *Erwachendes Denken: Geistige Leistungen aus evolutionsbiologischer Sicht*. Heidelberg: Spektrum Akademischer Verlag.

Köhler, W. 1921. *Intelligenzprüfungen an Menschenaffen*. 2nd ed. Berlin: Springer.

Kosel, G. 1957. *Produktivkraft Wissenschaft*. Berlin: Verlag Die Wirtschaft.

Kosing, A., and [for Author Collective]. 1968. *Die Wissenschaft von der Wissenschaft: philosophische Probleme der Wissenschaftstheorie*. Berlin: Dietz-Verlag.

Kriss-Rettenbeck, L., and M. Liedtke, eds. 1986. *Erziehungs- und Unterrichtsmethoden im historischen Wandel*. Bad Heilbrunn: Klinkhardt.

Krüger, L., ed. 1970. *Erkenntnisprobleme der Naturwissenschaften: Texte zur Einführung in die Philosophie der Wissenschaft*. Königsstein: Neue wissenschaftliche Bibliothek.

Kuhn, T. S. 1962. *The structure of scientific revolutions*. Chicago: University of Chicago Press.

Landau, B., and R. Jackendoff. 1993. "What" and "where" in spatial language and spatial cognition. *Behavioral and Brain Sciences* 16:217–265.

Landsberger, B. 1926. Die Eigenbegrifflichkeit der Babylonischen Welt. *Islamica* 2:355–372.

Langer, J. 1980. *The origins of logic: Six to twelve months*. New York: Academic Press.

Langer, J. 1986. *The origins of logic: One to two years*. New York: Academic Press.

Lefèvre, W. 1971. *Zum historischen Charakter und zur historischen Funktion der Methode bürgerlicher Soziologie*. Frankfurt am Main: Suhrkamp.

Lefèvre, W. 1978. *Naturtheorie und Produktionsweise*. Darmstadt: Luchterhand.

Lefèvre, W. 1981. Rechensteine und Sprache. In *Rechenstein, Experiment, Sprache: Historische Fallstudien zur Entstehung der exakten Wissenschaften*, ed. P. Damerow and W. Lefèvre, 115–168. Stuttgart: Klett-Cotta.

Leontjew, A. N. 1973. *Probleme der Entwicklung des Psychischen*. Frankfurt am Main: Athenäum Fischer.

Levin, I., and D. Zakay, eds. 1989. *Time and human cognition: A life-span perspective*. Amsterdam: North-Holland.

Levinson, S. C. 1991. *Relativity in spatial conception and description*, Working paper. Vol. 1. Nijmegen: Cognitive Anthropology Research Group at the Max Planck Institute for Psycholinguistics.

Levinson, S. C. 1992. Primer for the field investigation of spatial description and conception. *Pragmatics* 2:5–47.
Lévy-Bruhl, L. 1910. *Les fonctions mentales dans les sociétés inférieures*. Paris: Alcan.
Lévy-Bruhl, L. 1922. *La mentalité primitive*. Paris: Alcan.
Lewis-Williams, J. D., and T. A. Dowson. 1988. The signs of all times: Entopic phenomena in Upper Palaeolithic Art. *Current Anthropology* 29:201–245.
Liben, L. S., A. H. Patterson, and N. Newcombe, eds. 1981. *Spatial representation and behavior across the life span: Theory and application*. New York: Academic Press.
Lloyd, B., and J. Gay, eds. 1981. *Universals of human thought: Some African evidence*. Cambridge: Cambridge University Press.
Lonner, W. J. 1980. The search for psychological universals. In *Handbook of cross-cultural psychology*, ed. H. C. Triandis and A. Heron, vol. 1, 143–204. Boston: Allyn and Bacon.
Luria, A. K. 1971. Towards the problem of the historical nature of psychological processes. *International Journal of Psychology* 6:259–272.
Luria, A. K. 1976. *Cognitive development: Its cultural and social foundations*. Cambridge, MA: Harvard University Press.
Malinowski, B. 1966. *Coral gardens and their magic*. 2nd ed. London: Allen and Unwin.
Mandl, H., and H. Spada, eds. 1988. *Wissenspsychologie*. München: Psychologie Verlags Union.
McCloskey, M. 1983. Naive theories of motion. In *Mental models*, ed. D. Gentner and A. L. Stevens, 299–324. Hillsdale: Erlbaum.
McCloskey, M., and R. Kargon. 1988. The meaning and use of historical models in the study of intuitive physics. In *Ontogeny, phylogeny, and historical development*, ed. S. Strauss, 49–67. Norwood: Ablex.
McDermott, D. 1982. A temporal logic for reasoning about processes and plans. *Cognitive Science* 6:101–155.
Miller, A. I. 1984. *Imagery in scientific thought creating 20th-century physics*. Boston: Birkhäuser.
Minsky, M. L. 1975. A framework for representing knowledge. In *The psychology of computer vision*, ed. P. H. Winston, 211–277. New York: McGraw-Hill.
Minsky, M. L. 1985. *The society of mind*. New York: Simon & Schuster.
Nelson, T. O., ed. 1992. *Metacognition: Core readings*. Boston: Allyn and Bacon.
Newell, A., and H. A. Simon. 1972. *Human problem solving*. Englewood Cliffs: Prentice-Hall.
Nissen, H. J., P. Damerow, and R. K. Englund. 1991. *Frühe Schrift und Techniken der Wirtschaftsverwaltung im alten Vorderen Orient*. 2nd ed. Bad Salzdetfurth: Franzbecker.
Nissen, H. J., R. K. Englund, and P. unter Mitarbeit von Damerow. 1993. *Die Lexikalischen Listen der Archaischen Texte aus Uruk*. Berlin: Gebr. Mann.
Opwis, K. 1992. *Kognitive Modellierung: Zur Verwendung wisssensbasierter Systeme in der psychologischen Theoriebildung*. Bern: Huber.
Piaget, J. 1950. *Introduction à l'épistémologie génétique*. Paris: PUF.
Piaget, J. 1978. Die historische Entwicklung und die Psychogenese des Impetus-Begriffs. In *Die Psychologie des 20. Jahrhunderts, Bd. VII: Piaget und die Folgen: Entwicklungspsychologie, Denkpsychologie, Genetische Psychologie*, ed. Gerhard Steiner, 64–73. Zürich: Kindler.
Piaget, J., and R. Garcia. 1989. *Psychogenesis and the history of science*. New York: Columbia University Press.
Piattelli-Palmerini, M., ed. 1980. *Language and learning: The debate between Jean Piaget and Noam Chomsky*. London: Routledge.
Pick, H. L., and L. P. Acredolo, eds. 1983. *Spatial orientation: Theory, research, and application*. New York: Plenum.
Pinxten, R., ed. 1976. *Universalism versus relativism in language and thought*. The Hague: Mouton.
Pinxten, R., I. von Dooren, and F. Harvey. 1983. *Anthropology of space: Explorations into the natural philosophy and semantics of the Navajo*. Philadelphia: University of Pennsylvania Press.
Planck, M. 1936. *The philosophy of physics*, trans. W. Johnston. New York: Norton
Planck, M. 1949. *Vorträge und Erinnerungen*. 5th ed. Stuttgart: Hirzel.

Poortinga, Y. H. 1992. Towards a conceptualization of culture for psychology. In *Innovations in Cross-cultural psychology. Selected papers from the tenth international conference of the international association for cross-cultural psychology*, ed. S. Iwawaki, Y. Kashima, and K. Leung, 3–17. Amsterdam: Swets & Zeitlinger.

Renn, J. 1993. Einstein as a disciple of Galileo: A comparative study of concept development in physics. *Science in Context* 6:311–341.

Russell, B. 1931. *The scientific Outlook*. London: George Allen & Unwin.

Saxe, G. B., and J. K. Posner. 1983. The development of number cognition: Cross-cultural perspectives. In *The development of mathematical thinking*, ed. H. P. Ginsburg, 292–317. New York: Academic Press.

Schiefenhövel, W., and I. Bell-Krannhals. 1986. Wer teilt, hat teil an der Macht: Systeme der Yams-Vergabe auf den Trobriand Inseln, Papua-Neuguinea. *Mitteilungen der Anthropologischen Gesellschaft in Wien* 116:19–39.

Schmandt-Besserat, D. 1992. *Before writing*. Austin: University of Texas Press.

Scribner, S. 1975. Recall of classical syllogism: A cross-cultural investigation of error on logical problems. In *Reasoning: Representation and process in children and adults*, ed. R. J. Falmagne, 153–173. Hillsdale: Erlbaum.

Scribner, S., and M. Cole. 1981. *The psychology of literacy*. Cambridge, MA: Harvard University Press.

Sethe, K. 1939. *Vom Bilde zum Buchstaben: Die Entstehungsgeschichte der Schrift*. Leipzig: Hinrichs.

Shneiderman, B., and G. P. Kearsley. 1989. *Hypertext hands on! - An introduction to a new way of organizing and accessing information*. Reading, MA: Addison-Wesley.

Steele, A. D. 1965. Über die Rolle von Zirkel und Lineal in der griechischen Mathematik. In *Zur Geschichte der griechischen Mathematik*, ed. O. Becker, 146–202. Darmstadt: Wissenschaftliche Buchgesellschaft.

Stern, E. 1920. Probleme der Kulturpsychologie. *Zeitschrift für die gesamte Staatswissenschaft* 3:267–301.

Stevenson, R. J. 1993. *Language, thought and representation*. Chichester: Wiley.

Stichweh, R. 1984. *Zur Entstehung des modernen Systems wissenschaftlicher Disziplinen: Physik in Deutschland 1740-1890*. Frankfurt a. M: Suhrkamp.

Szabó, Á. 1968. *Anfänge der griechischen Mathematik*. Munich: Oldenbourg.

Thompson, J. E. S. 1960. *Maya hieroglyphic writing*. Norman: University of Oklahoma Press.

van de Vijver, F. J. R., and F. J. M. Hutschemaekers, eds. 1990. *The investigation of culture: Current issues in cultural psychology*. Tilburg: Tilburg University Press.

von Hohenzollern, J. G. P., and M. Liedtke, eds. 1988. *Naturwissenschaftlicher Unterricht und Wissenskumulation: Geschichtliche Entwicklung und gesellschaftliche Auswirkungen*. Bad Heilbrunn: Klinkhardt.

von Weizsäcker, C. F. 2011. *Die Verantwortung der Wissenschaft im Atomzeitalter (1957)*. Göttingen: Vandenhoeck & Ruprecht.

Vosniadou, S., and W. F. Brewer. 1987. Theories of knowledge restructuring in development. *Review of Educational Research* 57:51–67.

Vosniadou, S., and W. F. Brewer. 1992. Mental models of the earth: A study of conceptual change in childhood. *Cognitive Psychology* 24:535–585.

Vygotsky, L. 1962. *Thought and language*. Ed. and trans. E. Hanfmann & G. Vakar. Cambridge, MA: MIT Press. (Original of 1934: *Izbrannye psichologiceskie issledovanija: myslenie i rec*).

Wapner, S., and B. Kaplan, eds. 1983. *Toward a holistic developmental psychology*. Hillsdale: Erlbaum.

Weinert, F. E., and M. R. Waldmann. 1988. Wissensentwicklung und Wissenserwerb. In *Wissenspsychologie*, ed. H. Mandl and H. Spada, 161–199. Munich: Psychologie Verlags Union.

Weingart, P., ed. 1972. *Wissenschaftssoziologie I, Wissenschaftliche Entwicklung als sozialer Prozeß*. Frankfurt am Main: Athenäum-Fischer-Taschenbuch-Verlag.

Werner, H., and B. Kaplan. 1963. *Symbol formation: An organismic-developmental approach to language and the expression of thought*. New York: Wiley.

Wertheimer, M. 1920. *Über Schlussprozesse im produktiven Denken*. Berlin: Weltkreisverlag.

Wertheimer, M. 1925. Über das Denken der Naturvölker: Zahlen und Zahlgebilde. In *Drei Abhandlungen zur Gestalttheorie*, ed. M. Wertheimer, 106–163. Erlangen: Palm & Enke.

Wertheimer, M. 1945. *Productive thinking*. New York: Harper.

Wertheimer, M. 1957. *Produktives Denken*. Frankfurt am Main: Kramer.

Wertsch, J. V., ed. 1985. *Culture, communication, and cognition: Vygotskian perspectives*. Cambridge: Cambridge University Press.

Whorf, B. L. 1956. *Language, thought, and reality, selected writings of B. L. Whorf*, J. B. Carroll. Cambridge, MA: MIT Press.

Wiser, M. 1988. The differentiation of heat and temperature: History of science and novice-expert shift. In *Ontogeny, phylogeny, and historical development*, ed. S. Strauss, 28–48. Norwood: Ablex.

Wiser, M., and S. Carey. 1983. When heat and temperature were one. In *Mental models*, ed. D. Gentner and A. L. Stevens, 267–297. Hillsdale: Erlbaum.

Wolff, M. 1978. *Geschichte der Impetustheorie: Untersuchungen zum Ursprung der klassischen Mechanik*. Frankfurt am Main: Suhrkamp.

Wolkow, G. N. 1970. *Soziologie der Wissenschaft. Studien zur Erforschung von Wissenschaft und Technik*. Berlin: Dietz Verlag.

Wozniak, R. H., and K. W. Fischer, eds. 1993. *Development in context: Acting and thinking in specific environments*. Hillsdale: Erlbaum.

Wundt, W. 1900–1920. *Völkerpsychologie: Eine Untersuchung der Entwicklungsgesetze von Sprache, Mythus und Sitte*. Vol. 10 vols. Leipzig: Kröner.

Zilsel, E. (1976). *Die sozialen Ursprünge der neuzeitlichen Wissenschaft* (ed. W. Krohn). Frankfurt am Main: Suhrkamp.

GPSR Compliance

The European Union's (EU) General Product Safety Regulation (GPSR) is a set of rules that requires consumer products to be safe and our obligations to ensure this.

If you have any concerns about our products, you can contact us on ProductSafety@springernature.com

In case Publisher is established outside the EU, the EU authorized representative is:

Springer Nature Customer Service Center GmbH
Europaplatz 3
69115 Heidelberg, Germany

Batch number: 09457412

Printed by Printforce, the Netherlands